Willingness to Pay Framework

Willingness to Pay Framework

Climate Change Mitigation in Households

Dalia Štreimikienė
Asta Mikalauskienė

CRC Press
Taylor & Francis Group
Boca Raton London New York

CRC Press is an imprint of the
Taylor & Francis Group, an **informa** business

First edition published 2021
by CRC Press
6000 Broken Sound Parkway NW, Suite 300, Boca Raton, FL 33487-2742

and by CRC Press
2 Park Square, Milton Park, Abingdon, Oxon, OX14 4RN

© 2021 Taylor & Francis Group, LLC

CRC Press is an imprint of Taylor & Francis Group, LLC

This research was funded by a grant (No. S-MIP-17-131) from the Research Council of Lithuania.

Library of Congress Cataloging-in-Publication Data
Names: Štreimikienė, Dalia, author. | Mikalauskienė, Asta, author.
Title: Willingness to pay framework : climate change mitigation in
households / Dalia Streimikiene and Asta Mikalauskienė.
Description: First edition. | Boca Raton : CRC Press, 2021. |
Includes bibliographical references and index.
Identifiers: LCCN 2021021419 (print) | LCCN 2021021420 (ebook) |
ISBN 9780367643768 (hardback) | ISBN 9781003126171 (ebook)
Subjects: LCSH: Climate change mitigation–Social aspects–Case studies. |
Climate change mitigation–Social aspects–Lithuania. |
Willingness to pay–Case studies. | Willingness to pay–Lithuania.
Classification: LCC TD171.75 .S777 2021 (print) |
LCC TD171.75 (ebook) | DDC 363.738/746–dc23
LC record available at https://lccn.loc.gov/2021021419
LC ebook record available at https://lccn.loc.gov/2021021420

ISBN: 978-0-367-64376-8 (hbk)
ISBN: 978-0-367-64768-1 (pbk)
ISBN: 978-1-003-12617-1 (ebk)

Typeset in Palatino
by Newgen Publishing UK

Contents

Contents

Authors

Prof. dr. Dalia Štreimikienė is professor at Vilnius University, Kaunas faculty and Leading Research Associate at Lithuanian Energy Institute. Her main area of research includes sustainable development, sustainability assessment and corporate social responsibility. She is an author of more than 100 papers in international journals referred at WoS: H-index 19. She is invited Editor for several special issues of *Sustainability* journal and Editor-in-Chief of international journal *Transformations in Business and Economics* referred at WoS. Orcid: 0000-0002-3247-9912. https://www.researchgate.net/profile/Dalia_Streimikiene2

Prof. dr. Asta Mikalauskienė is professor at Vilnius University, Kaunas faculty. Her main area of research includes sustainable development, sustainability assessment and corporate social responsibility. She is an author of many papers in international journals referred at WoS. Orcid: https://orcid.org/0000-0002-4301-2058. https://www.researchgate.net/profile/Asta_Mikalauskiene

Authors

Acknowledgement

This research was funded by a grant (No. S-MIP-17-131) from the Research Council of Lithuania.

Introduction

There are two main ways to mitigate climate change in households related to energy consumption: increase of energy efficiency and the use of renewable energy sources. However, there are a number of market failures and other barriers that hamper the implementation of these climate change mitigation measures.

Various climate change mitigation policies and measures are designed in households to overcome or reduce many of the economic, social, psychological, technological and regulatory barriers of climate change mitigation, and only a detailed analysis of these barriers and the policies used to overcome them can identify the most effective and efficient policies.

In order to reduce the psychological and social barriers of climate change mitigation, it is essential to ensure that policies are tailored to the preferences of the population with these preferences assessed and measured using willingness to pay (WTP) methods. Assessing the WTP for climate change mitigation allows not only to determine the preferences of the population in climate change mitigation but also to assess the external economic benefits linked to climate change mitigation measures.

This monograph analyses WTP for climate change mitigation studies and presents results of several case studies conducted in Lithuania: study on the assessment of WTP for renewable micro-generation technologies in private houses and WTP for energy renovation of multi-flat buildings. These two measures have the greatest potential for climate change mitigation in Lithuanian households.

The identified preferences for climate change mitigation during WTP studies allow to overcome more effectively all barriers linked to climate change mitigation in households. These barriers are economic, social, information, regulatory and technological. However, there are also psychological barriers of climate change mitigation, otherwise known as the seven dragons of inaction. When formulating climate change mitigation policy, it is very important to take into account that public preferences are variable and that climate change mitigation policies can change these preferences to form new ones.

Therefore, the main goal of this monograph is to prepare a tool for the assessment of population preferences in climate change mitigation using the WTP approach and then applying this tool in Lithuania by assessing

1

WTP in two main areas that have the greatest potential for greenhouse gas emission reduction: installation of renewable micro-generation technologies in households and energy renovation of multi-flat buildings.

The analysis of climate change mitigation policies and measures in Estonia, Latvia and other EU Member States, based on the assessment of their effectiveness in removing barriers of climate change mitigation, allowed to provide recommendations for the development of innovative climate change mitigation measures.

The following objectives have been set to achieve the main goal of this study:

1. Analysis of the social benefits of climate change mitigation measures and their integration methods based on the assessment of public preferences;
2. Assessment of WTP for climate change mitigation measures in Lithuanian households;
3. Analysis of Lithuanian climate change mitigation policies and measures in households and evaluation of their effectiveness based on the achieved results;
4. Analysis of the effectiveness of global energy and climate change mitigation measures in overcoming the main barriers to climate change mitigation in households;
5. Development of innovative climate change mitigation measures in households, based on the conducted study of WTP for climate change mitigation measures in Lithuania and analysis of the effectiveness of climate change policy measures.

The monograph consists of five parts, in which the above-mentioned objectives are discussed and the conclusions of the conducted research are presented in the final part of the monograph.

1

Assessment of Climate Change Mitigation Benefits and Their Integration in Climate Policies

1.1 Insights of Behavioural Economics in Developing Energy and Climate Policies

Among the many challenges inherent in developing climate change policies, reconciling of immediate local costs with long-term globally distributed benefits must be one of the most difficult tasks. Not surprisingly, the political debate was focused on the former. Discussions on how and to what extent to address climate change have focused almost exclusively on reducing the costs of greenhouse gas (GHG) emissions. But the motivation to adopt climate change mitigation measures and cover certain costs linked with these actions must ultimately come from the idea that climate change mitigation provides benefits to society. These perceived benefits are determined by various elements. There are benefits that depend solely on the individual's situation. They also spread across borders and are uncertain. These benefits are linked to different perspectives on risk aversion, risk perception, time choice, intergenerational responsibility and ethical motivations. They differ in terms of magnitude and characteristics of avoidable damage in the future.

Energy consumption and energy-efficient investments are closely linked to consumer decision-making and behaviour. These aspects have led to a greater interest in developing policy interventions targeting energy demand and there is great interest among scholars and decision makers in assessing the changes in consumer behaviour in response to these interventions. Behavioural economics can provide new perspectives that can help shape climate change mitigation policies, based on insights on how individuals value opportunities, make decisions and change their behaviour.

It is important to stress that energy policy is not only about climate change mitigation but also how it is closely linked with the security of energy supply and energy affordability. Climate change mitigation policies also have to deal with the problems of expensive and interruptible renewable energy sources

(RES). If consumer behaviour can be changed to reduce energy demand or to make energy demand more responsive in time and space to weather-related energy supply shortages, this could make a significant contribution to facilitating the use of renewable energy by climate change mitigation policies. Conversely, failure to address public concerns about security of energy supply or climate change mitigation policies could undermine the achievement of ambitious GHG reduction targets (Pollitt & Shaorshadze, 2011).

Traditionally, the economics is focused on how price changes affect consumer behaviour. Research on behavioural economics and psychology has shown that non-monetary interventions are more conducive than monetary interventions in order to change consumer behaviour. It has also been shown that intelligently tailored non-monetary interventions can increase the impact of monetary interventions when used together. The current increased interest in behavioural economics research is linked to the needs of guiding decision aiding processes in such diverse areas as public health, finance, and law. Policy makers and researchers have increasingly recognized that the behavioural economics can help in decision-making and shaping of more effective policies in many sensitive areas. In order to save energy and reduce GHG missions the decision makers need to consider strategies targeting behavioural changes. Behavioural economics provides insights that can underpin these efforts.

Behaviours linked to household energy consumption and climate change mitigation cover these areas: energy consumption including curtailment behaviour, and habits; energy efficiency investments; and contribution to public goods such as renewables. These three issues of energy consumption and climate change mitigation are closely interrelated.

Supporting the use of RES and combating global warming are a public good. Behavioural economics can help to understand why and under what circumstances individuals would like to voluntarily contribute to these public goods. These contributions can be monetary (i.e. when individuals pay premiums for green energy) or non-monetary (i.e. when individuals operate in a sustainable way but sacrifice their comfort). Behavioural economics states that many individuals are "conditional co-workers" and that they value honesty. Individuals would like to contribute to public goods if they knew others are contributing as well. In essence, the difference between an explanation of 'behavioural change' and an explanation of 'traditional' behaviour is that, under the first option, individuals do not want to be free-riders; they simply care about others doing so, and consider this to be unfair; under the latter, individuals have an innate tendency for free riding, and this will be the case even if others contribute to climate change mitigation by paying taxes or taking actions. Moskovitz (1993) argued that customers would voluntarily sign up and pay higher electricity tariffs if the additional money raised were used to support renewable energy projects and environmental activities. Since then, energy utilities have offered green energy tariffs. These green tariffs are a contribution to the well-being of all society. If individuals really want to pay contributions voluntarily, public policy can take advantage of this trend by creating mechanisms that increase the likelihood of those contributions. The public economy has addressed the problem of under-supply of public goods through mechanisms such as taxes and points-of-supply mechanisms (PSM). Under PSM, individuals make voluntary contributions to a project, and are able to withdraw their contributions if the required reference amount is not collected.

Rose *et al*. (2002) in field experiments tested the PSM as a measure to collect means for a renewable energy program initiated by Niagara Mohawk Power Corporation in the US. During this experiment PSM increased the percentage of participation in the green energy program of control groups significantly compared with treatment groups. When PSM was tested in the field, the observed sign-up percentage was much higher than that of other green pricing programs that ask for voluntary contributions. One of the problems with green energy sales in countries like the UK is the complementarity. The specific amounts of RES set by the government mean that when buying green electricity, they simply allocate renewable energy that would in any case be available to a certain group of consumers. This is because all consumers will still have to pay for renewable energy, whether they pay a voluntary or conditional contribution.

Behavioural economics suggests that if people were motivated by the "warm glow" effect, then the provision of monetary incentives would reduce their motivation to contribute to public goods. If individuals pay a fine for behaviour that undermines public goods, their intrinsic motivation to avoid such behaviour may decrease. This "crowding-out hypothesis" is intellectually appealing, but there are no empirically tested results in energy field. Jacobsen *et al*. (2012) analysed the billing data from green electricity program participants and non-participants in Memphis, Tennessee. Jacobsen *et al*. (2012) estimated that households participating in programme had increased their electricity consumption by 2.5% after the inclusion of the plan. They explain this by the "buying in" mentality of these households. The paying for green electricity by participating in programme displaces their motivation to reduce their own electricity consumption. However, the impact was not large enough to offset the environmental benefits of paying for green electricity. Therefore, the net effect was a reduction of GHG emission and programme may be considered as successful. Also it was found that individuals can do more to deliver public goods if their contribution is "publicly" recognized. Thus, symbolic gifts such as Y. Clips, Mugs, and Stickers provided in exchange for an energy saving are one way to encourage energy savings (Yoeli, 2010).

When the increase of prices is being considered as socially or politically unacceptable, governments have sometimes addressed public calls to promote energy savings through the media. In the absence of price signals, the traditional economy would not expect public calls to have impacts on behavioural changes, as individuals would already have optimized their choices at a given price. However, the behavioural economics suggests that appealing to the public can raise awareness and encourage altruistic individuals to save more energy. In addition, some behavioural economists postulate that public calls can influence the changes of social norms.

When the externalities of economic activity are negative, the traditional economy suggests that the following measures can be used to reduce the extent of harmful behaviour: taxes, restrictions (i.e. quotas or prohibitions) and tradable certificates or quotas. Taxes will reduce harmful behaviour by increasing prices, and quotas will discourage behaviour by limiting the amount of "bads" that are allowed or available. Taxes levied on activities that cause negative externalities are called Pigou taxes and have traditionally been applied to goods such as cigarettes and alcohol or such environmentally destructive activities such as pollution.

Tradable quotas set a general permissible level of activity, and limited quotas for these activities are granted to entities (usually companies). These allowances are tradable, so those with the lowest marginal cost of reducing negative activity to acceptable levels will be the ones to sell them. The permits sold have been used to regulate fisheries and air pollution (Frey 2005; Tietenberg, 2003). The traditional economics assumes that the difference between these three policies is purely economic and administrative efficiency. Psychologists argue that another important difference between these instruments is that they differ as much as they send signals that displace intrinsic motivation.

Frey (1999) provides that the motivation of environmental activity is moral. He argues that both the sale of allowances and taxes will have two opposite effects on consumers: an increase in the cost of activity will discourage behaviour, but it will also reduce the internal morale known as "crowding out" effect. Frey (1999) argues that environmental morale will be reduced more by trading permits than by taxes: traded permits can be seen as similar to indulgences sold for sins in the Middle Ages (Goodin, 1994). These permits can give the impression that sin is acceptable as long as one pays the price for it. Frey also suggests that both low and high environmental taxes would be more effective than mid-level taxes. He argues that because of low taxes, consumers may feel that protecting the environment is something that needs to be done out of moral obligations. On the other hand, high environmental taxes make harmful behaviour too costly and dominate the crowding-out effect. Meanwhile, average environmental taxes displace internal motivation, but they are not enough to reduce behaviour due to external motivation. In climate policy, the negative action that needs to be reduced is GHG emissions.

Taxes, restrictions and tradable permits are often used as instruments of climate change mitigation policies. The European Union has implemented the largest Emissions Trading Scheme (EU ETS). Until now, tradable permits have been assigned to companies but not to end-users. On the other hand, consumers in many countries are directly paying carbon and energy taxes. Unfortunately, evidence to displace environmental morale because of the signalling effects of taxes and tradable permits has not been empirically evaluated, but research has shown that the phenomenon exists in laboratory and field experiments (Frey & Jegen 2001; Deci *et al.*, 1999). However, its relevance to actual consumer behaviour and the extent of the impact have not yet been actually determined. Meanwhile, public awareness of the impact of these measures is important for the political economy of public policy. As the public (especially environmental activists) realize that the taxes and tradable permits are morally lower because they seem to sanction pollution, it may be politically more difficult for the government to implement these mechanisms.

Measures of climate change mitigation in households worldwide can be divided into five main groups: mandatory provision of more detailed information on goods and services; development of environmentally friendly consumer attitudes; promotion of environmentally friendly (green) public procurement; adjustment of the tax system; and promoting the eco-efficiency of production and services (Pollitt & Shaorshadze, 2011).

The provision of more detailed information on the goods and services is aimed at providing the buyer, as an individual who makes the final consumption decisions (what to buy, how much to buy, etc.), not only general information about the quantity and quality of the product, country of origin, etc., but also information on the raw materials and substances used and the potential environmental impact of the product throughout its life cycle. Of particular importance is the development of environmentally friendly consumer attitudes through education and training institutions. According to global experience, the best results are achieved when such education begins at the kindergarten age of children.

Measures aimed solely at the individual attitudes and habits of the buyer often do not produce the expected results, which is why in recent years there has been an increased focus in the world on creating a favourable legal environment. One of the most effective such measures is the promotion of green public procurement.

Adjusting the tax system through green budget reform is another important administrative tool for promoting sustainable consumption and production, as well as climate change mitigation. One such option is the wider application of one of the most important principles of sustainable development – the polluter pays principle.

Promoting the eco-efficiency of production and services is an important means of reducing the environmental impact of not only production but also consumption. Therefore, the consumption of eco-efficient products and services addresses the key challenge of sustainable consumption, so that increasing consumption does not increase the negative impact on the environment, which is largely dependent on human behaviour.

A review of recent research has shown that climate change mitigation behaviours of households can crowd out public support for government climate change mitigation actions due to misperceptions about sufficient progress. This is called the crowding-out effect in public opinion. Surveys and experiments conducted in Japan have revealed this phenomenon after the closure of the Fukushima Daiichi Nuclear Power Plant. The effects of crowding out may have been due to a growing awareness of the importance of individual action and a consequent decline in support for the government's climate change mitigation policies. Studies in US have shown that, contrary to expectations, information on climate change has had no direct or indirect impact on climate change policy support, as the media often provide conflicting information about the causes and extent of climate change. Higher-income households are more supportive of climate change mitigation policies. Interestingly, political orientation was closely linked to policy support. In contrast to previous studies, the results showed that older adults expressed more support for climate change mitigation policies than younger individuals, and education had no significant effect (Gifford, 2011).

Various interventions to promote behavioural change are currently being criticized in the scientific literature for concerns that they may reduce support for comprehensive climate change mitigation policies necessary to address climate change. Academic debates describe the terminology of this phenomenon differently: the rebound effect, the negative impact of one activity on another, moral licencing or negative resistance, but there remains a common concern that successful interventions to persuade people to take appropriate mitigation action sometimes reduce the likelihood of other important climate change mitigation measures to be implemented. For example, reminding people about their past environmental and climate change mitigation actions or encouraging new ones can reduce public support for government climate change mitigation policies.

Another mechanism called moral licencing is that good behaviour can make a person feel fair and therefore give him or her the freedom to behave in less fair ways in the future. Researchers' insights have revealed that reminding a person of the energy-saving behaviour of the past strengthens people's perception of themselves as moral persons. Subsequent research has shown that manipulating that information to make people look even more moral and contribute to climate change mitigation has had no effect on their support for climate change mitigation policies. Surveys in the US have shown that when it comes to saving energy, if people prioritized the importance of individual behaviour over government action, it reduced their acceptance that energy and environmental issues should be a national priority. Thus, people are shifting their perception of where to take action on climate change mitigation, thus reducing the importance of the role of the state in climate change mitigation policies (Pollitt & Shaorshadze, 2011).

Research in the US and Europe has shown that while most people under-stand that climate change is an important issue, only a small proportion of people change their lifestyles and reduce GHG emissions. This is due to a variety of structural barriers, such as unfavourable infrastructure, but recent research shows that psychological barriers are much more important in deter-ring households from behaving appropriately to mitigate climate change (Gifford, 2011).

While many households are engaged in some climate change mitigation measures, most could do even more, but they are hampered by the seven identified categories of psychological barriers or 'inactivity dragons': limited knowledge of the problem, ideology, comparisons with other important people, behaviour change costs, negative attitudes towards experts and authorities, perceived risk of change and positive but not significant change in behaviour. To remove these barriers, psychologists need to work with other scientists, technical experts and policy makers. (Gifford *et al.*, 2011).

1.2   Willingness to Pay (WTP) Estimation Methods

Integrated assessment models, combined with physical climate model, can assess the future climate change policy's impact on climate change mitiga-tion. These models have been widely used to determine the benefits of cli-mate policy and by comparing the marginal benefits with marginal costs, the socially optimal level of climate stabilization measures can be defined. An alternative method of evaluating climate change mitigation benefits is to encourage individuals to evaluate these benefits by themselves. Determining the benefits in this way is difficult given the complex nature of global climate change phenomena. Economists, political scientists and policy analysts have begun to address this issue through a variety of ways to assess consumer willingness to pay (WTP) for increased climate stability.

Economists describe the benefits of environmental policy as a collective WTP to preserve a certain amount of environmental well-being (Stavins, 2007). Typically, researchers collect data on the WTP for public goods like clean air using one or more of the following methods: (1) calculating the dollar price based on what others actually pay for access to organic goods (the travel cost method); (2) identification of price differences between other-wise similar assets that differ only in that they have access to market goods (hedonic pricing); and (3) fuzzy estimation techniques (contingent valu-ation methods (CVMs)) that use surveys to elucidate hypothetical scenarios (O'Conner & Spash, 1999). Climate stability WTP studies are almost always based on the assessment of hypothetical scenarios and are therefore suitable for CVM studies. In addition, this method is particularly appropriate for our purposes because it reflects the value of both use and non-use ones (Stavins,

2007). This is important because the benefits of climate stability for voters in developed countries are largely theoretical and indirect.

The shortcomings of WTP methods for environmental valuation are linked to their tendency to reflect the larger challenges faced in applying cost-benefit analysis in environmental policies. On the one hand, due to the extreme uncertainty, long-term and social conflicts that characterize climate change, the monetary consequences of climate change become very severe (Conner, 1999). In addition, as the impact of climate transcends international boundaries, a comprehensive assessment must be made on a global scale. The actual assessment of climate stability also depends heavily on discount rates.

Studies of the social and behavioural aspects of energy consumption show that higher incomes and a more favourable lifestyle are associated with higher levels of WTP in energy-saving equipment (Lutzenhiser, 1993). Feedback on electricity use surveys among households has found significant negative effects of environmental awareness on energy consumption (Brandon & Lewis, 1999). This negative effect was even greater for households with a positive attitude towards the environment. A comprehensive review of the literature on the determinants of energy-related behaviour is beyond the scope of this analysis. However, these studies are useful in revealing the empirical basis for researching climate stability WTP studies in certain approach and socio-economic categories.

Scholars have highlighted the importance of contextual geographical differences and local opportunities in explaining environmental behaviour (Poortinga, 2004). Generational changes are also important for explaining the recent changes in residential energy consumption (O'Neill & Chen, 2002). Residential energy consumption is also determined by different psychological needs and attitudes related to different generations, ethnic groups and socio-economic groups. In addition, the American Association of Psychologists has recently identified a number of mental models that influence individuals' understanding of and response to climate change as well as their action (Gifford, 2011). For example, formulating a climate stabilization instrument according to its effects on the weather can lead to the feeling of chaos and helplessness. These insights are a starting point for determining plausible explanations for the variance of WTP estimates. The next section provides the main plausible explanations for the variance of WTP estimates.

A vast majority of WTP studies on climate stability have been conducted over the past decade, and each of the 27 studies included in this analysis was conducted after 1998. WTP assessment methods for climate change mitigation have emerged recently. To identify time trends, researchers have only recently begun to investigate the influence on climate change mitigation benefits of certain key explanatory variables such as uncertainty in climate outcomes (Cameron, 2005; Viscusi & Zeckhauser, 2006) and travel frequency (Brouwer *et al.*, 2008). Environmental goods assessed according to WTP estimates vary widely, ranging from climate stabilization policies in general (Cameron, 2005), to investments in green energy (Wiser, 2007; Hoyos & Longo, 2009), reduced temperature changes and food shortages due to rising gas prices (Viscusi & Zeckhauser, 2006; Solomon & Johnson, 2009) and specific carbon sequestration mechanisms (Brouwer *et al.*, 2008). The common vehicles used to calculate WTP are income taxes (Bohringer, 2004) and gasoline prices (Viscusi & Zeckhauser, 2006), increased energy prices (Berrens *et al.*, 2004) and higher household costs. The majority of studies applied CVM. Other methods include a probit model based on discrete routine responses to specific scenarios and extrapolation from public opinion polls (Bohringer, 2004). Researchers use several types of question in CVM studies. In the open-ended questions they simply ask respondents to "name their price" for a specific climate change mitigation measure. Such issues often occur with payment cards, indicating the range of possible prices the respondent is willing to pay (Solomon & Johnson, 2009). The questions are usually presented as single or multiple dichotomous choice questions that capture "yes"/"no" answers to a randomly selected WTP offer.

Important explanatory variables for WTP assessment of climate stability are gender (Viscusi & Zeckhauser, 2006), education (Berk & Fovell, 1999), level of perceived responsibility (Brouwer *et al.*, 2008), temperature rise (Berk & Fovell, 1999), the type of vehicle payable (Wiser, 2007), understanding the effects of climate change (Nomura & Akai, 2004), the efforts of respondents (Berrens *et al.*, 2004) and the uncertainty of climate change outcomes (Cameron, 2005).

The two main ways to mitigate climate change related to energy consumption in households are to increase energy efficiency and the use of RES. State climate change mitigation measures should be based on an assessment of their social benefits; therefore the population WTP for energy efficiency measures (energy-saving equipment, insulation and renovation of housing, modernization of heating and ventilation systems) and RES allows to assess the social benefits of climate change mitigation measures and propose policy instruments for the integration of social benefits and social costs.

Understanding and targeting the behavioural change in terms of private household energy consumption and conservation are essential in achieving energy efficiency and climate change mitigation targets (Stadelmann, 2017). However, this is a difficult task to take into account the 'Energy Efficiency Paradox' that provides a clear divergence between cost-effective energy-efficient choices and behaviour (Ramos *et al.*, 2015), emphasizing the 'irrational' issues of individual decision-making (Pollitt & Shaorshadze, 2011). Essential observations from behavioural economics show the systematic deviation between people's knowledge and attitudes and the assumptions of rational choice and maximization of utility in classical economics (Shogren & Taylor, 2008; Frederiks *et al.*, 2015; Pothitou *et al.*, 2016; Stadelmann, 2017).

Numerous studies examine the links between consumers' social and demographic characteristics, norms and attitudes, and energy efficiency habits and behavioural choices, including the choice of household appliances. For example, Yue *et al.* (2013) investigate in China's Jiangsu Province households WTP for different energy-saving behaviours, including appliance purchasing activities. Various hypotheses have been raised, such as that the energy saving behaviour is influenced by demographic factors, environmental awareness, knowledge, social norms and the prices of appliances. Reynolds *et al.* (2012) use several regression techniques, including Tobit modelling, to investigate consumer WTP for fluorescent lamps that are more efficient in St. Lucia. Hori *et al.* (2013) use three subgroups of independent variables, namely global warming awareness, environmental behaviour, and social interactions, and two demographic variables – age and income – to explain the energy-saving behaviour across a multiple regression model in five Asian towns. Jacobsen *et al.* (2012) analysed the impact of energy prices on consumers WTP for high-efficiency goods. Sardianou (2007) examines the determinants of energy-saving habits in Greek households by identifying the importance of several social, demographic and environmental awareness factors. Abrahamse *et al.* (2005) investigated the socio-demographic and psychological factors associated with energy consumption and savings in Dutch households, and found the huge importance of psychological factors. Many other studies were focused on the impact of various factors on WTP for different types of electric appliances when introducing energy efficiency labelling techniques (e.g., Shen & Saijo, 2009; Zhou & Bukenya, 2016; Ward *et al.*, 2011; Revelt & Train, 1998).

In order to assess the social benefits of climate change mitigation and identify the support schemes for renewables, the assessment of the population's WTP for public goods, such as RES, is widely used.

The number of studies published in the last year focusing on consumer preferences towards RES has increased steadily (Sundt & Rehdanz, 2015). All these studies vary widely based on the energy-related characteristics and geographical location; these studies analyse but also apply different WTP valuation techniques (Johnson *et al.*, 2010; Menegaki, 2008; Streimikienė & Mikalauskiene, 2014; Sundt & Rehdanz, 2015).

Menegaki (2008) has grouped all the studies of renewable energy evaluation into the following main methods, depending on the research field from where the research is launched: stated preference techniques, revealed preference techniques, financial option theory, emergy analysis and economic but not welfare-based oriented methods. Financial option theory–portfolio analysis valuates projects in line with their anticipated risks and returns. These approaches valuate renewables not on the basis of their stand-alone cost, but on the basis of their overall portfolio cost with expected portfolio risk. Emergy analysis methods are applied mainly in economical engineering for assessing the net value of environmental projects to human society. And other economic but not welfare-based oriented methods are "various other economic methods and techniques which do not fall under above groups and are not welfare-based either" (Menegaki, 2008), whereas the first two methods, stated and revealed preferences, are intended for energy consumers' WTP for renewable energy and their technologies evaluation.

As can be seen in Figure 1.1, three main techniques are available in the field of WTP valuation: stated preference techniques, revealed preference techniques and conjoint analysis (CA).

Stated preference and revealed preference techniques are based on random utility theory. Only recently Louviere *et al.* (2010) demonstrated that CA

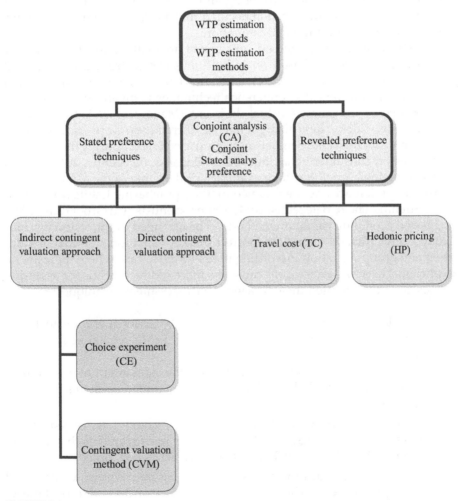

FIGURE 1.1

Methods applied in WTP studies. Source: Created by author based on Streimikiene *et al.* (2019).

doesn't belong to random utility theory and evolved out of the theory of "Conjoint measurement", which is "purely mathematical and concerned with the behaviour of number systems and not the behaviour of humans or human preferences". Sometimes scholars are confused regarding CA and choice experiment (CE) methods and they often are considering CA as a special case of CE (sometimes vice versa), which is considered as a stated preference technique. According to Louviere *et al.* (2010) though both the CE and CA methods apply experimental design to assess WTP, however, one of the main differences between them is that CA "methods depend on orthogonal arrays of attribute level combinations as ways to sample profiles from full factorial arrays of attribute levels", while stated preference methods do not have this limitation. Thus, considering all the controversy related to the CA method, the author of dissertation decided that a CA is not the most reliable method while performing WTP estimation.

The stated preference method, originated in mathematical psychology (Acito & Jain, 1980), is based on individual's choice from a hypothetical choice set (Adamowicz *et al.*, 1994) or his direct answer, whereas information for the analysis of revealed preference techniques is given out by markets, as produced by consumers' actual decisions (Menegaki, 2008; Štreimikienė & Baležentis, 2014). Revealed preference techniques were pioneered by American economist Paul Samuelson and are used for comparing the influence of policies on consumer behaviour (Štreimikienė & Baležentis, 2014). Travel cost and hedonic pricing can be examples of these methods, assuming that consumer preferences can be revealed by their purchasing habits (Menegaki, 2008; Štreimikienė & Baležentis, 2014). Other scientists (De Groot *et al.*, 2002) suggested economic valuation methods to be arranged in the following four groups: direct market valuation, indirect market valuation, contingent valuation and group valuation. According to scientists, in case where there are no explicit markets for services, one must use more indirect means for assessing values. Scientists pointed the following WTP assessment techniques:

1. Avoided cost assessment provides for the assessment of cost that would have been incurred in the absence of climate change mitigation measures;
2. Replacement cost assessment allows for the assessment of costs linked to services that could be replaced with human-made systems instead of natural;
3. Factor income provides the assessment of income that ecosystem services can provide like natural water quality improvements that increase commercial fisheries catch and thereby fishermen income.
4. Travel cost provides the assessment of ecosystem service value based on required travel. These costs are reflection of the value of the ecosystem service like recreation areas that attract distant visitors who are willing to pay to travel to it.

5. Hedonic pricing is based on the idea that prices people will pay for environmental benefits of associated goods, e.g. housing prices at beaches, will exceed the prices of identical inland homes with less attractive landscape.

One of the first comparative applications of the stated preference approach was made by Adamowicz *et al.* (1994). Scholars applied a stated preference model and a revealed preference model in their study while combining them both. The results provided that both, i.e. "hypothetical" stated preference and "actual behavior" revealed preference, techniques "provides evidence that the underlying preferences are in fact similar". Banfi *et al.* (2008) have stressed that though revealed and stated preference methods are used to evaluate WTP, the stated preference method is the preferred one, namely CE. This is because of the small size of energy-efficient houses' market and because the stated preference method made it possible to compare the WTP of people who have already experienced the additional comfort benefits of energy-saving measures with those who do not have such information. Other scientists agreed to this idea as well. Claudy *et al.* (2011) highlighted that many methods can be applied to estimate WTP; however, if only a small number of target households surveyed group exist, then it becomes very difficult to apply a revealed preference method; therefore, stated preference methods are more feasible in such conditions.

While comparing stated and revealed preference techniques it can be confirmed that stated preference techniques are more suitable for WTP for renewable energy resources (Štreimikienė & Ališauskaitė-Šeškienė, 2014). Revealed preference techniques require a well-developed market for green energy (Herbes *et al.*, 2015; Štreimikienė & Ališauskaitė-Šeškienė, 2014). Since climate change is a global phenomenon, both travel cost and hedonic pricing methods cannot assess the benefits of global climate change mitigation in a specific area (Štreimikienė & Ališauskaitė-Šeškienė, 2014).

Stated preference techniques are based on the idea that consumers are interested in the modes that energy is produced not just in energy supply (Štreimikienė & Baležentis, 2014). There are two main methods highlighted by Wood *et al.* (1995): direct and indirect CVM.

Direct CVM is like asking survey respondents directly how much value they place on a given good. However, such questions are biased, as respondents may have an incentive to either overvalue or undervalue their true WTP, depending on how the questions is formulated. In order to minimize some of the above-mentioned biases, Wood *et al.* (1995) developed an indirect approach. Wood *et al.* (1995) stressed that "indirect approach is more effective than asking direct WTP questions because goods or products are described as a collection of attributes and respondents must carefully weigh the trade-offs between attributes". Wood *et al.* (1995) also added, "because the product is not explicitly identified, and respondents are asked to state their preferences for attribute level, respondents' incentive to over- or under-report their true WTP minimize".

The CVM studies are asking each respondent a closed form question, whether they would accept to pay a given amount to obtain a given change in their status quo, by "yes" or "no" type answers (Christiaensen & Sarris, 2007). This method is often used for assessing public attitudes towards the use of RES for energy generation (Stigka *et al.*, 2014); therefore, it is used for the estimation of consumers' WTP for RES by providing the choice among various alternative RES like solar, wind, hydro or geothermal (Menegaki, 2008).

CE or discrete CE was initially developed by Louviere and Hensher *et al.* (1982). In CE respondents are asked to select their preferred alternative from a given set of choices and are typically asked to perform a sequence of such choices, i.e. respondents make trade-offs between all attributes across each alternative and are expected to choose their most preferred alternative (Campbell *et al.*, 2008). CE confronts "respondents with multiple questions in the following form: do you prefer A or B, where A and B are described by the level of the characteristics of a good or service" (van Putten *et al.*, 2014). Thus, the method of CE is well suited to the elicitation of trade-offs between different RES technology types described in the questionnaire (Kosenius & Ollikainen, 2013).

According to scientists (Boxall *et al.*, 1996), though CVM and CE approaches require respondents to state their preferences for environmental goods, the main differences can be obtained from the values derived from these two methods. Boxall *et al.* (1996) highlighted that the CE method has some advantages over the CVMs because of a less biased questionnaire form. In the CE situation, results "rely less on the accuracy and completeness of any particular description of the good or service, but more on the accuracy and completeness of the characteristics and features used to describe the situation". Therefore, the experimental nature of CE makes its advantage (Boxall *et al.*, 1996).

There are many valuation methods and survey types. Wood *et al.* (1995) analysed WTP among several key customer segments including residential and found significant differences. Hanley and Nevin (1999) applied WTP to estimate "of either an individual's WTP for an improvement in the quality or quantify of some environmental good". Roe *et al.* (2001) conducted the survey to elicit consumer's WTP for changes in environmental characteristics of residential electricity supply. Ek (2005) investigated customers attitudes towards electricity produced at wind farms.

Bergmann *et al.* (2006) applied the CE to assess public preferences over environmental and social impacts of hydro power plants, wind mills and biomass boilers in Scotland. Borchers *et al.* (2007) performed a contingent CE study for assessing public preferences and WTP for voluntary participation in green electricity programs. Banfi *et al.* (2008) applied the CE approach to estimate customer's WTP for energy-saving measures in residential buildings in Switzerland. Bergmann *et al.* (2008) used the CE method for analysis and comparison of preferences of urban and rural residents. Longo *et al.* (2008) studied WTP of energy consumers for different characteristics of energy programs that promote RES and applied CE. Longo *et al.* (2008) proved that stated preference studies on WTP energy security usually focus on short-term security of supply (black-outs), rather than on price volatility linked to long-term security of supply.

Zografakis *et al.* (2010) conducted a CVM study for evaluation of public acceptance of RES in Crete. Zorić and Hrovatin (2012) assessed WTP for renewable electricity in Slovenia. For assessment of consumer preferences Guo *et al.* (2014) estimated WTP for renewable electricity and used the CVM in Beijing, China. Štreimikienė and Baležentis (2014) conducted a study for assessment of WTP for RES in Lithuanian households by applying the CE method. Akcura (2015) assessed public preferences and WTP for renewables under a mandatory scheme in Japan.

In Table 1.1 the WTP studies on WTP for climate change mitigation measures linked to RES.

As can be noticed from Table 1.1, the majority of scholars in their WTP studies used the CE method or CVM; however, just a few of them (Scarpa & Willis, 2010; Claudy *et al.*, 2011; Štreimikienė & Baležentis, 2014; Oberst & Madlener, 2014; Lee & Heo, 2016) analysed WTP for microgeneration renewable technologies that can be installed in households. This is due to the fact that microgeneration technologies in most European countries are not widely spread though there are many policies targeting these technologies.

Claudy *et al.* (2011) indicated that home owners' WTP for microgeneration technologies "is significantly lower than the actual market prices of these technologies". The studies on social acceptance of renewable energy technologies were linked mainly to large renewable technology projects (Sauter & Watson, 2007); however, microgeneration technologies based on renewables have huge potential (Van der Veen *et al.*, 2009). Sauter and Watson (2007) noted that renewable microgeneration technologies applied in households do not only impact individuals' environment, e.g. noise or spoiling the landscape, but also require the WTP to use them.

The main microgeneration technologies applied in households are mentioned below (Scarpa & Willis, 2010; Willis *et al.*, 2011):

6. Solar photovoltaic technology is based on solar roof panels comprising thin layers of semiconductor material aimed to convert sunlight into electrical energy.
7. In micro-wind technology a turbine usually constructed on the roof converts kinetic energy of wind into electrical energy.
8. Solar thermal is a roof-based shade structure to absorb solar by energy-mounted collectors (panels), heat water or other fluids, and can power solar cooling systems as well. Solar thermal systems differ from PV systems that generate electricity rather than heat.

TABLE 1.1

Summary of WTP for renewable energy source studies

Study	Method of WTP assessment	Area of analysis	Renewable energy source	Region	Model
1. Wood *et al.* (1995)	CE	Residential, commercial and industrial	Renewable electricity	US	Probit model
2. Hanley and Nevin (1999)	Direct CV	Remote community	Wind farm, small-scale hydro and biomass	Scotland	None (stated WTP)
3. Roe *et al.* (2001)	CE	Households	Residential electricity	US	Linear model
	Hedonic pricing	Households	Residential electricity services	US	Linear ordinary least squares method
4. Nomura and Akai (2004)	Direct CV	Residents in large cities	Photovoltaic and wind power	Japan	None (stated WTP)
5. Ek (2005)	CE	House owners	Hydro biomass solar and wind	Sweden	Probit model
6. Borchers *et al.* (2007)	CE	Residential sector	Different renewable energy programs	US	Non-linear probability model
7. Bergmann *et al.* (2008)	CE	Rural and urban residents	Hydro, wind biomass	Scotland	Logit model
8. Longo *et al.* (2008)	CE	Households	Hypothetical RES programmes	UK	Random utility model
9. Banfi *et al.* (2008)	CE	House owners and apartment tenants	Energy saving	Switzerland	Logit model
10. Bollino, (2009)	CVM	Households	Renewable electricity	Italy	Probit model
11. Zografakis *et al.* (2010)	CVM	Households	RES project s	Greece	Logistic regression
12. Scarpa and Willis (2010)	CE	Households	Photo-voltaic, wind solar thermal, heat pumps, biomass	UK	Logit model
13. Claudy *et al.* (2011)	CVM	Households	Wind, solar panels, solar thermal, biomass	Ireland	Probit model

14.	Zorić and Hrovatin (2012)	CE	Households	Renewable electricity	Slovenia	Tobit model
15.	Aravena *et al.* (2012)	CVM	Households	Renewable electricity	Chile	Discrete choice random utility model
16.	Kosenius and Ollikainen (2013)	CE	Households	Wind, hydro biomass	Finland	Logit model
17.	Guo *et al.* (2014)	CVM	Households	Renewable electricity	China	Logit model
18.	Bigerna and Polinori (2014)	CVM	Households	Renewable electricity	Italy	Logistic regression
19.	Štreimikienė and Baležentis (2014)	Direct contingent valuation	Households	Renewable electricity	Lithuania	Non-parametric regression
20.	Oberst and Madlener (2014)	CE	Households	Wind, solar, biomass	Germany	Logit model
21.	Akcura (2015)	CVM	Households	Renewable electricity	UK	Probit model
22.	Chan *et al.* (2015)	CVM	Households	Renewable electricity	South Africa	Tobit model
23.	Dagher and Harajli (2015)	CVM	Households	Renewable electricity	Lebanon	Tobit model
24.	Grilli *et al.* (2015)	CVM	Households	Hydro, biomass electricity	Italy	Tobit model
25.	Yamamoto (2015)	Direct contingent valuation	Households with PV systems	Photovoltaic	Japan	None (stated WTP)
26.	Jung *et al.* (2015)	CVM	Households	Renewable energy	South Korea	Regression model
27.	Morita and Managi (2015)	CA	Households	Solar and wind power	Japan	Logit model
28.	Sun *et al.* (2015)	CVM	Households	Smog mitigation	China	Probit model and interval regression
29.	Vecchiato and Tempesta, (2015)	CE	Households	Renewable energy	Italy	Logit model
30.	Lee and Heo (2016)	CVM	Households	Solar and wind power	South Korea	Logistic regression

Source: Created by authors.

9. Biomass boilers and pellet stoves are usually microgeneration technologies that use wood chips or pellets and are applied for space heating and hot water supply needs.

10. Small-scale CHP technologies are combined heat and power generation systems with electrical power less than 200 kW (Streimikiene & Baležentis, 2013). Although technically CHP is not a "renewable" energy source, it is assigned to renewable technology because of its potential to save significant amounts of energy and reduce GHG emissions (Claudy *et al.*, 2010).

Microgeneration CHPs are grouped into mini-CHP and micro-CHP. Mini-CHP has installed capacity from few kilowatts up to 100 kW, whereas micro-CHP usually has installed capacity of around 10 kW of the installed electrical power capacity (Streimikiene & Balezentis, 2013).

Several small-scale CHPs compete on the market (Alanne & Saari, 2004; De Paepe *et al.*, 2006; Streimikiene & Balezentis, 2013):

1. Reciprocating engines – a power plant based on a reciprocating engine consists of a reciprocating engine (diesel, gas or multiple fuel) and a generator linked to the engine. However, they are noisy and not very attractive alternative for residential applications.

2. Stirling engines is also a reciprocating engine, but contrary to conventional diesel and gas engines its cylinder is closed and combustion takes place outside the cylinder. Furthermore, they have lower noise production and can be used in residential buildings. However, their low efficiency supports their use as backup power supply rather than one in continuous use.

3. Fuel cells produce electricity electrochemically, by combining hydrogen and atmospheric oxygen. Not only this technology has a very low emission rate, it is also noiseless, reliable and modular. However, the costs of a fuel cell plant can be up to three times higher than reciprocating engines making it an important drawback.

4. Micro-turbines are gas turbines with electrical power generation capacity from 25 kW to 250 kW. Micro-turbine plants produce low noise and are of small size; they are also more environmentally friendly than reciprocating engines; yet their electrical efficiency is low and they are expensive.

Claudy *et al.* (2010) highlighted in their study conducted in 2001 that all renewable energy technologies will have an increasingly important role in the future, as they provide a great potential for GHG emissions, reduce fossil fuel dependency and stabilize energy prices. Microgeneration technologies in particular have the potential to contribute favourably to energy supply security (Allen *et al.*, 2008); furthermore it could fundamentally change the relationship between energy companies and consumers (Watson, 2004) by literally turning the system upside down: as at least partial shift would be performed from an electricity system based on central power plants (like nuclear, coal or big natural gas-based power plants) to small-scale power generation at the domestic level (Sauter & Watson, 2007). In that case, consumers would become energy suppliers in their own right; however, a pre-condition for this change is the diffusion of microgeneration technologies into the market which will depend on consumers' acceptance of microgeneration technologies (Watson, 2004), i.e. their WTP for renewable energy microgeneration technologies.

Unfortunately, despite major marketing and public policy efforts, the diffusion of these particular technologies in most European countries is slow; thus, microgeneration technologies can be referred as resistant innovations – they face slow take up times as they require consumers to alter their existing belief structures, attitudes, traditions or entrenched routines significantly (Claudy *et al.*, 2010; Garcia *et al.*,,, 2007). Furthermore, the deployment of renewable energy in the residential sector also depends on consumers' intentions to adopt a technological innovation (Sardianou & Genoudi, 2013). And while classical economic theory suggests that individuals make consumption decisions that maximize their welfare given the capital constrained derived demand function, demand for one good or service occurs as a result of the demand for another intermediate/final good or service, and thus, consumers usually think of themselves as a central actor in a decision-making process (Lancaster, 1966; Sardianou & Genoudi, 2013).

It can be stressed that understanding of consumer preferences and WTP for renewable energy technology or microgeneration technology becomes even more important because of public policies promoting further the use of RES due to the targets set for renewables in final energy consumption, final electricity and heat consumption (Borchers *et al.*, 2007). Microgeneration technology is still in its growing stage in the market and public opinion towards it is crucial which has been highlighted by Sauter and Watson (2007), Wüstenhagen *et al.*, (2007), Allen *et al.* (2008), Van der Veen *et al.* (2009), Willis *et al.* (2011), and Sardianou and Genoudi (2013).

Knowing more about customer preferences in green energy is very important as well as having information about the foundations of these attitudes (Ek, 2005). In many cases results of WTP studies vary widely, which, according to Kraeusel and Möst (2012), "is the result of different methodologies and intermittent preferences of customers" (Kraeusel & Möst, 2012; Poortinga *et al.*, 2003). And although most of the existing research generally supports that individuals are willing to pay extra for RES (Akcura, 2015; Bigerna & Polinori, 2014; Borchers *et al.*, 2007; Nomura & Akai, 2004), zero WTP or negative WTP may exist as well.

Negative WTP indicates that respondents should be compensated in order to choose to use a product with a particular attribute (James *et al.*, 2009), while zero WTP means the respondent does not have to be paid for using such products nor is he willing to sacrifice to procure a good. However, some scientists exclude negative WTP, although that may lead to erroneous conclusions about the net social benefits of the proposed change (Hanley *et al.*, 2009). Furthermore, in the survey, it would be more exact to leave that econometric estimation of possible negative WTP, as it testifies the low or no interest of households (Christiaensen & Sarris, 2007) and "consequently, if there is not sufficient consumer willingness to pay, public funding is needed to support RES development" (Bigerna & Polinori, 2014). Thus, negative WTP can contribute to government's decision about the size of compensation in order to encourage consumers to use RES technologies. However, "if consumers take into account the environmental issues and consider that promoting RES will mitigate environmental damage, they are likely to attach a positive value to RES" and positive consumers thinking towards RES technologies may influence their WTP by augmenting the premiums they are willing to pay for such new technology (Bigerna & Polinori, 2014). Thus, the need for public funding might be reduced over time (Bigerna & Polinori, 2014).

WTP provides the assessment of public preferences for climate change mitigation policies and actions, especially in localized settings (Johnson *et al.*, 2012). Environmental awareness and beliefs are the main common explanatory elements in WTP studies, while others include income, education and political views (Johnson *et al.*, 2012). Although not all the above-mentioned scholars have examined and specified determinants or socio-demographic factors in their study, which affected their WTP assessments the most, the main determinants of WTP are summarized in Table 1.2.

Furthermore, when choosing between two methods, used by the majority of scientists, CE and CVM, in order to estimate consumers WTP for microgeneration technology, attention should be drawn that CVM is considered rather as a relict now – a while ago environmental valuation studies have been dominated by the CVM, which lasted for almost 20 years, nowadays CE is dominating in this area (Navrud & Bråten, 2007). And while applying a stated preference CE method, attributes of microgeneration technology play a crucial role. Summary of attributes provided in WTP studies are presented in Table 1.3.

1.3 Conclusions

As the analysis of the scientific literature has shown, the main socio-demographic characteristics that determine the WTP for RES and energy efficiency measures in households identified in empirical studies (Johnson & Nemet, 2010; Tol, 2013; Ma *et al.*, 2015) are age, gender, level of education, income, price, geographic location, position held and environmental awareness. Only a few empirical studies have found that belonging to

TABLE 1.2

Socio-demographic characteristics having impact on WTP for climate change mitigation

Determinants Study	Health impacts	Environmental impacts	Political views	Gender	Marital status	Education	Age	Vacancies	Income	Equipment use restrictions	Price	Contract terms	Race	Environmental organization affiliation	Geographical place
Wood et al. (1995)	+	+						+		+					
Roe et al. (2001)		+		+		+	+		+		+	+	+	+	+
Ek (2005)						+	+		+					+	
Borchers et al. (2007)				+			+				+				
Bergmann et al. (2008)		+				+	+	+	+		+				
Longo et al. (2008)				+	+	+	+		+		+	+		+	
Bollino (2009)				+		+	+	+	+						+
Claudy et al. (2011)				+			+								+
Zorić and Hrovatin (2012)		+		+		+	+		+		+				+
Aravena et al. (2012)				+			+		+		+			+	
Koseniu Ollikainen (2013)		+		+			+	+	+						+
Guo et al. (2014)				+		+	+		+					+	
Bigerna and Polinori (2014)				+		+	+		+						

(continued)

TABLE 1.2 (Continued)

Socio-demographic characteristics having impact on WTP for climate change mitigation

Determinants Study	Health impacts	Environmental impacts	Political views	Gender	Marital status	Education	Age	Vacancies	Income	Equipment use restrictions	Price	Contract terms	Race	Environmental organization affiliation	Geographical place
Streimikiene and Balezentis (2014)				+		+	+	+	+						
Oberst and Madlener (2014)				+	+		+	+	+		+				
Akcura (2015)				+			+		+					+	
Chan et al. (2015)		+		+			+		+						
Dagher and Harajli (2015)			+	+	+	+	+	+	+			+		+	
Grilli et al. (2015)		+		+			+		+					+	
Yamamoto (2015)				+		+	+		+						
Jung et al. (2015)				+		+	+								
Morita and Managi (2015)				+		+	+								+
Sun et al. (2015)							+		+						
Vecchiato and Tempesta (2015)				+			+								+
Lee and Heo (2016)				+		+	+		+						

Source: Created by authors based on Streimikiene *et al.* (2019).

environmental organizations, race, political views, expected health effects have an impact on WTP for renewable energy in households.

The vast majority of researchers for the assessment of WTP for renewable energy applied CEs or CVM; however just a few of them (Scarpa & Willis, 2010; Claudy *et al.*, 2011; Zhang *et al.*, 2019; Su *et al.*, 2018; Streimikiene & Balezentis, 2014; Oberst & Madlener, 2014; Lee & Heo, 2016) assessed the WTP for microgeneration technologies using RES in households. Most of the empirical research on WTP for RES sought to assess WTP for an increase in the share of RES in the electricity balance or for renewable energy projects in specific areas.

Although the meta-analysis studies did not include negative and zero estimates of WTP for climate change mitigation, some of the studies analysed yielded negative estimates of WTP for RES. Negative WTP estimates reveal that households expect compensation for the use of RES and the installation of microgeneration technologies in their homes (Bigerna & Polinori, 2014). Meanwhile, a zero assessment of WTP shows that residents are not willing to pay for renewable energy and do not want to receive compensation for the use of renewable energy in their homes. However, some researchers in their studies who received negative estimates of WTP did not include them in their econometric models, which distorted their estimates of the benefits of RES (Hanley *et al.*, 2009).

According to some researchers, negative estimates of WTP for RES indicate a low public interest in RES and microgeneration technologies (Christiaensen & Sarris, 2007), while others argue that negative WTP estimates require state support for these technologies to ensure their penetration in the market (Bigerna & Polinori, 2014).

Thus, even in the case of negative WTP for renewables, some policy recommendations can be generated regarding the necessary support for renewables. However, some researchers believe that negative estimates of WTP for renewables are due to consumers' lack of understanding of the environmental benefits of renewables, which, if they were aware of the benefits, would inevitably demonstrate a positive WTP for renewables. As behavioural economists have shown, media could make a significant contribution to shaping consumer preferences (Kahneman & Tversky, 1979). The positive opinion formed by the media about RES will definitely lead to higher WTP for these technologies (Zhang *et al.*, 2019). In addition, in the future, as the population becomes more concerned about the environment and about the benefits of RES (Bigerna and Polinori, 2014), the need for state support for these energy production technologies will decrease.

A thorough analysis of scientific literature led to the conclusion that studies on WTP for climate change mitigation in the energy sector are dominated by studies dealing with WTP for RES in households. At the same time there is scarce scientific literature on WTP for energy efficiency improvements in households, although energy savings ensure not only external benefits such as the use of RES but also direct cost savings for households due to lower

TABLE 1.3

Attributes used in empirical studies on WTP for climate change mitigation

Attributes Study	Capital cost	Installation cost	Investment risk	Mainte- nance cost	Payback period	Net electricity cost (or monthly electricity bill)	Annual energy saving	Recom- mended by	CO_2 reduction (contri- bution to climate protection)
Borchers et al. (2007)						+	+		
Bergmann et al. (2008)						+			
Longo et al. (2008)						+			+
Scarpa and Willis (2010)	+			+		+	+	+	
Kosenius and Ollikainen (2013)						+			+
Oberst and Madlener (2014)		+	+		+	+			+
Vecchiato and Tempesta (2015)						+			

Source: Created by authors based on Streimikiene *et al.* (2019).

energy bills. It is important to note that most studies of WTP for energy efficiency have focused on the impact of financial initiatives, subsidies and soft
loans on WTP for energy efficiency measures such as energy renovation of
residential buildings (Banfi *et al.*, 2008; Collins & Curtis, 2018; Portnov *et al.*,
2018; Liu *et al.*, 2018; Leung, 2018), but these empirical studies yielded rather
contradictory results on the impact of grants and soft loans on household
WTP for improving energy efficiency in their homes by renovating residential buildings. Banfi *et al.* (2008) in their study revealed that the size of
the price or premium that residents are willing to pay for green or passive
buildings is strongly negatively influenced by short-term financial initiatives
taken by the government. In their study, researchers (Banfi *et al.*, 2008) found
that various financial initiatives (tax cuts and loan incentives) result in
lower WTP for energy renovation. This negative impact of financial support

Degree of electricity self-supply	Social impacts (e.g. create new local employment)	Contract length	Air pollution	Landscape impact	Wildlife impact	Annual length of electricity shortages	Certification of origin	Minimum distance from houses	Size of the installed device	Inconvenience of system
	+		+	+	+					
	+					+				
		+								+
	+				+					
+	+									
							+	+	+	

measures can be explained using the insights of behavioural economics that material incentives to promote desired behavioural change (energy saving) have negative impact on desired behavioural changes (Pollitt & Shaorshadze, 2011; Gifford, 2011; Newell & Siikamkj, 2013).

In order to select and implement effective measures for the implementation of climate change mitigation policies, it is necessary to identify the preferences of the population and their WTP for the main policies and measures to mitigate climate change in households: the use of RES and energy efficiency improvements in households. The use of RES in households is primarily related to microgeneration technologies, and the most important means of increasing energy efficiency with the greatest potential for energy savings and GHG emission reduction in households is energy renovation of residential buildings.

2

Assessment of WTP for Climate Change Mitigation in Lithuania

2.1 Assessment of WTP for Renewables in Households

The willingness to pay (WTP) study begins with a definition of the problem to be solved. The example of the problem is how much customers are willing to pay for a specific product or for the improvement of a specific product. Thus, the first issue which is necessary to take into account in the WTP survey is concerned with the delineation of the drivers that respondents have to provide about their choice. As incentive is linked to a real or hypothetical product, the appropriate method should be derived based on the information available in the market. The conjoint choice analysis (CCA) is one of the most popular and widely accepted methods for defining WTP. Thus, this approach should be the first choice in a WTP study.

Despite the advantages, there are situations where CCA is not appropriate. The CCA uses different profiles of products to determine consumer preferences, and when the products have many attributes, the profiles to be assessed for these products become quite complex. If the mental effort of the respondents increases, then the analysis of their answers may lead to erroneous conclusions. In general, indirect methods outperform the direct ones. However, if there are no prevailing prices in the market, for example in the case of public goods, an alternative method is recommended.

Depending on the method chosen, an appropriate test plan for the determination of WTP should be developed. The study design depends on the type of measurement of the method, i.e. direct or indirect, and the composition of the indirect methods. Thus, taking into account the specific characteristics of the method, respondents are asked to evaluate the alternatives on the basis of specific questionnaires. Thus, the design, presentation and evaluation tasks, or the form of question, differ in each survey. Thus, a special WTP set-up procedure is established for each survey design.

The sample of respondents in the survey must be representative. It is necessary to ensure that demographics of the sample is matching the demographics of the target group. It should also be checked whether the public good is well known, as respondents who do not know enough about alternatives cannot make real statements. The minimum sample for this method is 20 respondents. There are two possible ways to collect necessary data: printed version of the questionnaire for the respondents and online version of questionnaires. Data collection is a very important step, especially in hypothetical methods, as they have to reconcile real purchase scenarios. If there are not enough respondents for the analysis, the results will also not be statistically verified.

After data collection, the WTP of the main decision-making stimulus can be defined. However, each assessment method involves different procedures for defining WTP. The main difference in setting a WTP is the type of measurement. With direct measurement methods such as CV, WTP can be determined immediately or respondents are asked directly whether they would pay the specific price if they bought a hypothetical public good. The price which is accepted by the respondent for the purchase of public goods is that respondent's WTP.

During indirect measurement methods WTP is obtained by asking respondents to carry out an assessment task. During CCA respondents are asked to evaluate pairs of individual alternative characteristics, and price is one of them. The benefits listed are then evaluated on the basis of these evaluation scores and are formulated to obtain the overall assessment of the product profile, which may be full or partial. On the other hand, when performing decomposition methods, respondents evaluate all product profiles. The specified provisions are then evaluated against these assessment scores and broken down to give individual values for each alternative characteristic. The definition of "alternative characteristic" refers to the levels of attributes that were used to form the profile of a particular alternative. By applying the revealed preference methods, the product profile is defined by a predetermined number of specific attributes. In addition, each attribute consists of a limited number of levels. A common feature of both the compositional and decomposition techniques is that they both give scores according to individual preferences for each level of the trait. These scores are named as utilities and are used to determine WTP. These utility values are calculated according to the composition of the method.

The assessment of WTP is based on the theory of random utility (Train, 2009). Let's treat $j = 1, 2, \ldots, n$ as the index for decision or policy makers and $i = 1, 2, \ldots, m$ as the index for the options being assessed. Every policy maker provides a certain probability to the ith option, U_{ij}. Option i is more preferable than the i'-th one if utility $U_{ij} > U_{i'j}, i \neq i'$; yet the latter values are not presented. Instead, the attributes of each option, x_{ij}, and policy maker, s_j, are observed. Thus, the representative utility or welfare is delineated as a function of the observed variables, i.e. $V_{ij} = V\left(x_{ij}, s_j\right)$. Taking into consideration the fact that policy analyst cannot assess all the factors affecting the welfare, a random error ε_{ij} allows to assess the disturbances due to unenclosed factors so that $U_{ij} = V_{ij} + \varepsilon_{ij}$. Therefore, the probability of selecting the ith option by the jth policy maker is assessed as follows:

$$P_{ij} = \Pr\left(U_{ij} > U_{i'j}, i \neq i'\right)$$

$$= \Pr\left(V_{ij} + \varepsilon_{ij} > U_{ij} + \varepsilon_{i'j}, i \neq i'\right)$$

$$= \int_{\varepsilon} I(\varepsilon_{i'j} - \varepsilon_{ij} < V_{ij} - V_{ij}, i \neq i')f(\varepsilon_j)d\varepsilon_j \qquad (1)$$

where $I(.)$ is the indicator function and $f(.)$ is the density function. Considering the welfare is linear in parameters β, one can provide this model: $U_{ij} = \beta x_{ij} + \varepsilon_{ij}$. The logit model to assess the probability of selecting the ith option by the jth policy maker is presented in the following way:

$$P_{ij} = \frac{exp(\beta x_{ij})}{\sum_{i=1}^{m} exp(\beta x_{ij})} \qquad (2)$$

To evaluate, in turn, for varying preferences and tastes of the policy makers, coefficients β can be arranged to vary across groups of various policy makers for capturing differences in preferences. This is called the mixed logit model (Revelt & Train, 1998; McFadden & Train, 2000). Assuming that β_j is the random vector of regression coefficients and $f(\beta_j | \theta)$ is the underlying density function with parameter vector θ. Therefore, for the mixed logit model, the probability of selecting the ith option by the jth policy maker can be presented as

$$P_{ij} = \int \frac{exp(\beta x_{ij})}{\sum_{i=1}^{m} exp(\beta x_{ij})} f(\beta | \theta)d\beta \qquad (3)$$

As each policy maker can encounter few experiments indexed over $t = 1, 2, \ldots, T$, a panel structure can be applied. Assume that y_{ijt} is equal unity if policy maker j selects option i during the tth experiment and is zero otherwise. So, the probability of revealing a specific pattern of choices is as follows (Train, 2009):

$$S_j = \int \prod_{t=1}^{T} \prod_{i=1}^{m} \left(\frac{exp(\beta x_{ij})}{\sum_{i=1}^{m} (\beta x_{ij})} \right)^{y_{ijt}} f(\beta | \theta)d\beta \qquad (4)$$

Parameters θ are assessed during the simulated procedure of maximum likelihood, which aims to maximize the given log-likelihood function:

$$SLL = \sum_{j=1}^{n} \ln\left[\frac{1}{R} \sum_{r=1}^{R} \prod_{t=1}^{T} \prod_{i=1}^{m} \left(\frac{exp\left(\beta_j^r x_{ij}\right)}{\sum_{i=1}^{m} exp\left(\beta_j^r x_{i'j}\right)} \right)^{y_{ijt}} \right] \qquad (5)$$

where $r = 1, 2, ..., R$ the index is drawn from $f\left(\beta_j \mid \theta\right)$. Then the procedure can be addressed in line with Hole (2007).

The obtained coefficients of the mixed logit model can be applied to obtain the assessment of WTP for certain features of the options. Considering that a (fixed) cost variable is embedded in the model, the WTP can be assessed as follows:

$$E\left(WTP_k\right) = -\frac{E\left(\beta^k\right)}{\beta^c} \qquad (6)$$

where k represents the kth attribute and β^c is the coefficient related with a cost variable.

$$E\left(W^*\right) = -\frac{E(\beta) x^*}{\beta^c}$$

It is possible to apply this model for assessing the change in utility linked to the choice of a certain option presented as a set of attribute values x^*. Taking into account the preferences of respondents shown by regression coefficients as well as the reference option obtained by attribute values x^0, it is possible to assess the change in utility based on the ratio of the difference in welfare over the negative of the coefficient (Bergmann *et al.*, 2008; Bennett & Blamey, 2001):

$$E\left(\Delta W^{*,0}\right) = \frac{E(\beta) x^0 - E(\beta) x^*}{\beta^c} \qquad (7)$$

where β is the coefficients vector in the mixed logit model.

In order to perform WTP for renewable energy technologies, the following specific micro-generation technologies were chosen, which are most popular in Lithuanian households: biomass, solar thermal, solar PV and micro-wind technologies. An unlabelled single-choice experiment with two common alternatives was developed. Attributes were set based on similar studies conducted in Japan (Nomura & Akai, 2004) and EU Member States (Bergmann *et al.*, 2006; Scarpa & Willis, 2010; Zografakis *et al.*, 2010).

The following characteristics were chosen for the current study: (i) technology installation costs, (ii) monthly energy bill, (iii) length of the warranty period, (iv) system inconveniences, and (v) degree of technology sharing. Each of the selected characteristics has four levels. In terms of system inconveniences, the four levels correspond to the specific circumstances related to the different energy production technologies, i.e. climatic conditions, the necessity for additional fuel, loud noise during operation phase and none of these. The additional characteristic like sharing was selected which reflects the population's preferences for the sharing of renewable energy technologies. In Table 2.1 the available options for each attribute are presented.

It should be noted that the environmental effect is not integrated in the model because, according to the pilot survey, this was not being considered by households as an important factor in the case of Lithuania. In fact, this can be explained by the fact that in Lithuania there is no serious environmental pressure related to energy consumption and climate change mitigation.

It is necessary to stress that some attributes (e.g. maintenance, user-friendliness, system flexibility) were not directly included in the survey because they were too difficult to assess when working with inexperienced users. The questionnaires were distributed to individual homeowners without having experience in micro-generation energy technology usage. However, the characteristics provided in Table 2.1 are indirectly linked to maintenance and flexibility criteria (Su *et al.*, 2018) and account for 40 alternatives, reflecting the total factor plan of 54 alternatives. There were two questionnaires distributed, each providing with 20 alternative choices (Wheeler, 2008). Respondents were asked to select one of two alternatives. Ten selection experiments were included in each questionnaire. Besides that, the respondents were free to select the status quo option. An example of a discrete choice experiment provided in the survey is given in Table 2.2.

The questionnaires were distributed and filled in by the respondents in June 2018. The respondents consisted of individual homeowners living in Kaunas District. Important condition for the respondents' selection was the requirement of no experience in application of micro-generation technologies

TABLE 2.1

Attributes and their levels applied during the discrete choice experiment

Attributes	1 level	2 level	3 level	4 level
Micro-generation technology installation costs, €	1,500	3,000	4,500	6,500
1. Monthly energy bill, €/month	16	30	35	38
2. Length of the warranty period, years	2	5	10	13
3. Complexity of requirements for operation of installation	Weather	Fuel	Noise	None
4. Degree of possibility for sharing technology	Very low	Low	Moderate	High

Source: Created by authors based on Su *et al.* (2018).

TABLE 2.2

The discrete choice experiment used in the survey

Attribute	Alternative A	Alternative B	Status quo
Technology installation costs, €	4,500	6,500	
Monthly average energy bill during all year, €/month	30	55	
Warranty period, years	2	10	
Availability of special requirements for technology operation	None	Special fuel required	
Possibility for sharing the technology for energy generation and consumption	Very low	Very low	
Please answer the following questions:			
1. Which of the two alternatives is more preferable for you based on information provided about each alternative?	☐ A	☐ B	
2. If you were allowed to choose among alternatives A and B and the status quo or baseline scenario without installation of new technologies, which option would you select?	☐ A	☐ B	☐ Base scenario

Source: Created by authors based on Su *et al.* (2018).

at home before survey. The sample size was calculated according to the Paniott formula, which required a sample of at least 99 respondents. During the experiment 104 respondents completed the questionnaires and conducted a total of 1,040 selection experiments.

Homeowners were provided with sets of combinations of certain micro-generation technology attributes and encouraged to select one more preferred alternative from the two sets. Continuing this, repeated choices by homeowners from alternative sets revealed the trade-offs they are ready to make between attributes and hence between four different micro-generation technologies.

The survey that was conducted allowed to find out the choice of the most preferable renewable energy sources (RES) and their distribution (Su *et al.*, 2018). A mixed logit model with an opt-out option was used to account for the differences in taste among respondents. Installation costs were selected as a fixed parameter, ensuring in this way greater model stability and a reasonable WTP estimate. All the remaining parameters were integrated as random, considering normal distribution. Installation costs, monthly bill and warranty period were included in the model as continuous variables, and performance requirements and degree of sharing were used as dummy variables taking into account very low possibility of sharing as base level and no specific requirements for the installation of technologies as base level. In Table 2.3, results of the mixed logit model are provided.

TABLE 2.3

Results of the mixed logit model

Variable	Coefficient	Standard error	z-value	Sig.
Mean				
Costs	-0.00104	0.000143	-7.27	***
Energy bill	-0.09366	0.015	-6.24	***
Warranty	0.250422	0.056134	4.46	***
Req1	-1.26372	0.387759	-3.26	***
Req2	-1.77965	0.280422	-6.35	***
Req3	-3.0829	0.586121	-5.26	***
Share2	0.176686	0.361369	0.49	
Share3	-0.11903	0.380056	-0.31	
Share4	0.242231	0.325592	0.74	
Standard deviation				
Energy bill	-0.00782	0.021144	-0.37	
Warranty	0.217705	0.048313	4.51	***
Req1	0.293656	0.536862	0.55	
Req2	0.068448	0.794067	0.09	
Req3	-1.63493	0.707709	-2.31	**
Share2	-0.37285	0.411201	-0.91	
Share3	0.402816	0.644225	0.63	
Share4	1.655886	0.36495	4.54	***
	N	1,402		
	LR $\chi^2(8)$	45.86		
	p-value	0.000		
	LL	-311.723		

Notes: (i) *** (**) denotes significant coefficients at the level of significance of 1% (5%); (ii) Req1 to Req3 correspond to the first three levels of requirements for operation in Table 2.1; (iii) Share2 to Share4 correspond to the last three levels of degree of possibility for sharing in Table 2.1.

Source: Created by authors based on Su *et al.* (2018).

On the basis of the suggestions of the previous literature (Scarpa & Willis, 2010; Samuelson & Zeckhauser, 1988), a mixed logit model with an opt-out option was applied that provided the dependent variable to be zero for both alternatives in the study (Table 2.3).

Probability ratio tests confirmed that the values of the standard deviations were zero; therefore it was confirmed that respondents' tastes vary greatly in terms of the impact of decision variables. Further manipulation of the distribution of parameters did not make significant changes in the evaluation results.

We have evaluated two models: one with an opt-out option and the other without an opt-out option (the left and right sections of Table 2.3). The models draw the same outcome regarding the influence of the determinants of WTP.

In Table 2.3 the significant coefficients have plausible signs: negative coefficients in addition to installation costs show that households are less likely to select alternatives associated with higher investment needs. The same is true for the monthly energy bill, but the same absolute increase in the monthly energy bill has a greater impact in comparison with the cost of installation.

An extended warranty period has a positive effect on the likelihood of selecting the appropriate alternative. Consequently, the availability of additional performance requirements reduces the attractiveness of the alternative for respondents.

In addition, noise appeared to be the most undesirable characteristic of technology. Given the change in taste, the monthly energy bill showed an insignificant coefficient, so respondents had no difference between the changes in monthly energy bills in their choice of energy technology. For all remaining variables, differences in preferences were noticed at the level of at least one of each characteristic.

The opportunity of sharing technology proved to be insignificant. So, households did not consider this to be a criterion for choosing an energy technology. It can be explained by certain factors related to the Lithuanian situation. Negative experiences of collectivization often inhibit collaborative initiatives in households. However, the reasons why respondents are not willing to share micro-generation technologies are worth further investigation.

The integration of monetary variables in the mixed logit model provided for monetary values of attributes of micro-generation technologies (Table 2.4). Increasing the average monthly energy bill by €1 reduces the installation costs by an average of €90 to keep the same level of utility for the respondents. This number reflects the expected timing of the transaction and the discount rate. The growth for the one-year guarantee period is worth of €241. Respondents

TABLE 2.4

WTP estimates for different attributes

	Energy bill (€)	Warranty period (years)	Requirements for technology operation			Opportunity of sharing technology		
			Weather	Fuel	Noise	Low	Moderate	High
$E(WTP_k)$	-90	241	-1,219	-1,716	-2,973	170	-115	234
CI	-107	161	-1,919	-2,315	-4,144	-507	-827	-368
	-73	322	-518	-1,117	-1,802	848	598	836

Notes: (i) *CI* stands for the 95% CI based on the delta method; (ii) requirements for operation are compared to case of no requirements; (iii) degrees of possibility for sharing are compared to very low possibility.

Source: Created by authors based on Su *et al*. (2018).

wanted to pay €2,973 to avoid the noise associated with the operation of micro-generation technologies. The confidence intervals (CI) for these traits showed a significant WTP values and estimates for sharing zero values. This again shows that respondents did not give a clear value for sharing energy technologies.

The four defined micro-generation technologies were subsequently assessed in the light of welfare changes. Using Equation (7), biomass, solar and wind micro-generation technologies were compared with solar PV.

The last rows in Table 2.5 present changes in welfare associated with the replacement of solar technology by other micro-generation technologies measured in monetary terms.

Therefore, it can be stressed that solar PV micro-generation technologies have emerged as the most desirable technology for Lithuanian households. Replacing PV micro-generation technologies with any other technology would make a negative change in customers' welfare. Also, micro-wind technology proved to be the least attractive option, valued at about €6,000 lower than the basic version of solar PV technology. In fact, much of this change can be linked to the noise exposure during the operation of micro-wind technology. In addition, the operation of biomass boilers requires certain fuels, so the population value micro-wind technology less compared with solar PV technology.

In Lithuania, a negative WTP was obtained for certain renewable energy, micro-generation technologies. However, the survey showed that Lithuanian households are willing to pay extra only for the guarantees of the chosen technology, which consists of solar photovoltaic, biomass boilers, solar thermal and micro-wind. For the use of solar thermal and micro-wind technologies households want to be compensated for choosing that technology with a certain characteristic. In any case, these results should not be interpreted directly but should be treated as an alternative classification tool based on the monetary value of the selected features. It is necessary to highlight that other models based on different attributes can provide different WTP values for the same technology.

In scientific literature the negative WTP estimates are obtained as well. A negative WTP shows that respondents should be paid to select a product

TABLE 2.5

Changes in welfare linked to four micro-generation technologies

Micro-generation technology	Solar thermal	Biomass boiler	Solar PV	Micro-wind
$E(WTP)$, €	1,363	-507	3,363	-2,597
$E(\Delta W^{\text{ solar PV}})$, €	-2,001	-3,870		-5,960
$E(WTP)$, €/month	15	-6	37	-29
$E(\Delta W^{\text{ solar PV}})$, €/month	-22	-43		-66

Source: Created by authors based on Su *et al.* (2018).

(James *et al.*, 2009), and a zero WTP indicates that the respondent does not want to pay for the use of such products, nor does want to be compensated. Although in some cases the results associated with a negative WTP may be ignored; this may provide for mistaken conclusions about the net social benefits of the renewable energy technologies (Hanley *et al.*, 2009). Therefore, taking negative WTP into account may be beneficial in that it indicates low or no household interest in renewable energy (Christiaensen & Sarris, 2007). In the absence of a strong interest in using particular technology (based on negative WTP estimates), the extent of public support to promote RES can be estimated on the basis of a WTP estimate (Bigerna & Polinori, 2014)

Consequently, a negative WTP estimate may also provide some contributions to the government's decision on the amount of compensation necessary to stimulate consumers to use renewable energy technologies. Nevertheless, the integration of an environmental dimension in the model can also change the results and provide that a positive WTP will be obtained for all micro-generation technologies. The shift to a more positive approach to RES technologies by households and changes in public preferences may have an impact on the WTP, particularly the bonuses that can be provided to ensure the introduction of new energy technologies (Bigerna & Polinori, 2014). Therefore, the amount of public funding for advancement in the use of RES may decline over time (Bigerna & Polinori, 2014). Therefore, environmental and other socio-economic features may be taken into account in WTP assessments. In addition, the analysis should be repeated for different user groups and at different time frames.

2.2   Assessment of WTP for Renovation of Multi-Apartment Buildings

According to the Lithuanian National Energy Independence Strategy (2018) more than 70% of multi-apartment buildings available in Lithuania are energy inefficient. Taking into account the fact that about 70% of the households in Lithuania live in multi-apartment buildings and about 100% of these buildings were built before 1993, this makes a huge energy saving and GHG emission reduction potential in Lithuania. Therefore, the assessment of energy renovation barriers in multi-apartment buildings and households' WTP for energy renovation is a very important task in order to develop policies and measures that are able to overcome the main barriers of energy renovation in households. Since 2013 in Lithuania about 2,000 apartment buildings were renovated. The average rate of energy renovation is 500 buildings per year. There are about 35,500 multi-apartment buildings currently in Lithuania that require energy renovation. Due to such slow pace it will be possible only to renovate all these buildings in 65 years.

Thus, as the slow pace of energy renovation is not satisfying the needs of policy makers and residents of these buildings, it is necessary to define the main barriers of the energy renovation process and to provide solutions to overcome these barriers and ensure a faster energy renovation process in Lithuania. The performed assessment of WTP for energy renovation in Lithuania would help to deliver more insights into the problem and develop policies and measures (Streimikiene & Balezentis, 2020).

There are main research questions to be solved during the case study on the assessment of WTP for energy renovation of multi-flat buildings in Lithuania, which are provided below:

1. What are the main organizational barriers of energy renovation in multi-apartment buildings?
2. What are the main economic barriers of energy renovation in multi-apartment buildings?
3. Is governmental support provided for energy renovation creates initiatives for energy renovation of multi-apartment buildings?
4. What are the most preferable ways for residents to pay for energy renovation of multi-flat buildings?
5. What is WTP for energy renovation by paying for it based on monthly energy bills?
6. What is WTP for energy renovation by paying the total costs of renovation before it starts?
7. Are residents of multi-flat buildings willing to share the costs of energy renovation with lower income households to faster renovation processes?

The study of WTP assessment was performed in September 2019 by applying a standardized telephone survey (N = 104). The survey included 39 multiple choice questions. The questionnaire consisted of several sections addressing specific issues. In the first section of the questionnaire, respondents were asked to provide their demographic data (Table 2.6); in the second section of the questionnaire, questions about the housing conditions of respondents were provided (Table 2.7); in the third section of the questionnaire, the main drivers of energy renovation were disclosed (Table 2.8); in the fourth section of the questionnaire, the average monthly energy costs of renovation were revealed (Table 2.9); in the fifth section of the questionnaire, the main barriers for energy renovation were identified by asking specific questions (Table 2.10); in the sixth section the WTP for energy renovation was disclosed (Table 2.11) and in the last section questions about the willingness to share energy renovation costs with low-income households living in the same multi-flat building were provided (Table 2.12). The survey consisted of multiple choice questions.

TABLE 2.6

Demographic data of sample

Gender	Frequency	%
Female	48	46.2
Male	56	53.8
Marital status of respondents		
Married	72	69.2
Divorced	11	10.6
Single	11	10.6
Widow	3	2.9
Cohabitants	7	6.7
Age of respondents, years		
Up to 23	3	2.9
23–34	19	18.3
35–44	43	41.3
45–65	34	32.7
Above 65	5	4.8
Employment status of respondents		
Unemployed	8	7.7
Retired	4	3.8
Employed in the public sector	37	35.6
Employed in the private sector	46	44.2
Self-employed (business)	9	8.7
Income of households per month, €		
Less than 300 €	10	9.6
300–500 €	13	12.5
501–1,000 €	36	34.6
1,001–1,500 €	24	23.1
1,501–2,000 €	15	14.4
2,001–3,000 €	3	2.9
Above 3,000 €	3	2.9
Education attained by respondents		
Secondary	10	9.6
Higher education	14	13.5
Tertiary education	80	76.9

Source: Created by authors based on Streimikiene and Balezentis (2020).

During the survey more than 1,000 phone calls were made. The response rate was 10%. The professional polling agency was hired to perform this survey. The targeted respondents were households living in multi-flat, non-renovated apartments built before 1993. Hierarchical sampling was applied to obtain a representative sample from ten Lithuanian regions. The respondents were also assessed based on the age (lower threshold was 18 years of age) taking into account the responsibility of being the head of the household. Demographic data of the respondents in the sample is provided in Table 2.6.

TABLE 2.7

Housing condition of respondents

Indicator	Frequency	%
Size of household		
One member	5	4.8
Two members	33	31.7
Three members	29	27.9
Four members	31	29.8
Five members	6	5.8
Type of building construction		
Constructed from bricks	12	11.5
Constructed from blocks	92	88.5
Construction year of the building, year		
Before 1961	13	12.5
1962–1993	90	86.5
Since 2010	1	1.0
Size of apartment in multi-flat buildings, m²		
Bellow 20	4	3.8
20-50	40	38.5
51-70	40	38.5
71-100	20	19.2
Number of rooms in apartment, rooms		
1	2	1.9
2	7	6.7
3	52	50.0
4	42	40.4
5	1	1.0
Number of floors in building, floors		
4	30	28.8
5	53	51.0
9	21	20.2
Apartment ownership status		
Private	73	70,2
Rented	31	29.8
Decision making on renovation		
Making decision alone	25	24.0
Making decision in consultation with members of household	79	76.0
Housing association in multi-flat building		
Available	46	44.2
Not available	56	53.8
Don't know	2	1.9
Energy labelling class of the building		
C	1	1.0
D	10	9.6
Don't know	93	89.4

Source: Created by authors based on Streimikiene and Balezentis (2020).

TABLE 2.8

Main drivers of energy renovation of multi-flat buildings

Indicator	Frequency	%
Average energy bill for electricity during the cold season		
0–20 €	4	3,8
21–50 €	60	57,7
51–100 €	32	30,8
101–200 €	4	3,8
201–300 €	1	1,0
301–400 €	3	2,9
Average energy bill for electricity during the warm season		
0–20 €	16	15,4
21–50 €	53	51,0
51–100 €	31	29,8
101–200 €	4	3,8
Average energy bill for heating during the cold season		
0–20 €	2	1,9
21–50 €	7	6,7
51–100 €	34	32,7
101–200 €	52	50,0
201–300 €	8	7,7
301–400 €	1	1,0
Average energy bill for heating during warm season		
0–20 €	53	51,0
21–50 €	40	38,5
51–100 €	8	7,7
101–200 €	2	1,9
201–300 €	1	1.0
Satisfaction of respondents with heating comfort of apartment		
No	94	90.4
Yes	10	9.6
Satisfactions of respondents with heating expenditures and their matching to heating comfort		
No	104	100.0
Consideration of energy renovation option		
Yes	95	91.3
No	9	8.7
Possession of information about available state support for energy renovation		
Information is known	104	100.0
Adequacy of state support for energy renovation		
Not adequate	104	100.0
Encouragement by state support available to renovation apartment		
Yes	1	1.0
No	89	85.6
Don't know	14	13.5

TABLE 2.8 (Continued)

Main drivers of energy renovation of multi-flat buildings

Indicator	Frequency	%
Necessity to renovate multi-apartment building		
Yes	100	96.2
No	2	1.9
Don't know	2	1.9
Partially energy renovation performed in multi-flat apartment building		
Not performed	58	55.8
Windows were changed in the apartment and building	10	9.6
Doors were changed in building	27	26.0
Implemented roof insulation of building	9	8.7

Source: Created by authors based on Streimikiene and Balezentis (2020).

TABLE 2.9

Average monthly energy cost per m² for average size of apartment of 53.3 m²

Indicator	€/month	€/m²
Average energy bill for electricity during the cold season	62	1.16
Average energy bill for electricity during the warm season	48	0.90
Average energy bill for heating during the cold season	125	2.35
Average energy bill for heating during the warm season	30	0.56
Average energy bill for electricity and heating during the entire year	152	2.85

Source: Created by authors based on Streimikiene and Balezentis (2020).

TABLE 2.10

Main barriers of energy renovation in multi-flat buildings

Indicator	Frequency	%
The main barriers preventing energy renovation		
High costs of energy renovation	2	1.9
Low income of households	7	6.7
Absence of households' association in multi-flat buildings	57	54.8
Reluctance to take a loan which is necessary for energy renovation	35	33.7
Lack of information about possibilities of energy renovation	2	1.9
Preparedness to sell apartment and reluctancy to invest in energy renovation due to this reason	1	1.0
Importance of availability of responsible organization able to solve of all organizational issues of energy renovation		
High importance	97	93.3
No importance	5	4.8
Don't know	2	1.9

Source: Created by authors based on Streimikiene and Balezentis (2020).

TABLE 2.11

Willingness to pay for energy renovation of multi-flat buildings

Indicator	Frequency	%
WTP for renovation per month in case of 40% energy savings achieved after energy renovation of multi-apartment buildings		
Bellow 10 €	48	46.2
11–50 €	44	42.3
51–100 €	12	11.5
WTP for renovation before starting it in case of 40% of energy saving achieved after energy renovation of multi-apartment buildings		
Below 1000 €	5	4.8
1,001–2,000 €	10	9.6
2,001–3,000 €	64	61.5
3,001–4,000 €	10	9.6
4,001–5,000 €	13	12.5
More than 5,000 €	2	1.9
WTP for renovation if costs are included in monthly energy bill in case of 40% energy savings achieved after energy renovation of multi-apartment buildings		
0%	2	1.9
Up to 10% rise in monthly energy bills	5	4.8
10–20% rise in monthly energy bills	3	2.9
21–30% rise in monthly energy bills	6	5.8
31–40% rise in monthly energy bills	28	26.9
41–50% rise in monthly energy bills	16	15.4
51–60% rise in monthly energy bills	21	20.2
61–70% rise in monthly energy bills	15	14.4
71–80% rise in monthly energy bills	3	2.9
91–100% rise in monthly energy bills	1	1.0
Above 100% rise in monthly energy bills	4	3.8
Preferences of payment for renovation		
Total sum to be paid before starting energy renovation	10	9.6
Payment is based on increased monthly energy bills	94	90.4

Source: Created by authors based on Streimikiene and Balezentis (2020).

As can be noticed from Table 2.6, 46% of the respondents were females and 53.8% were males. Majority (70%) of the respondents were married. Over 40% of the respondents were 35–44 years old. Just approximately 8% of the respondents were unemployed. Almost 4% of the respondents were retired. Respondents employed in the private sector were 44.2% while in the public sector were 35.6%. More than 34% of the respondents had a monthly income in the range from €500 to €1,000. Just approximately 6% of the respondents had incomes higher than €2,000 per month. And about 10% of the respondents had incomes below €300 per month. At the same time, 14% of the respondents had incomes between €1,500 and €2,000 per month. Almost 80% of the respondents had a tertiary education.

TABLE 2.12

Willingness to share renovation costs with lower income neighbours

Indicator	Frequency	%
Availability of households unable to pay for energy renovation of their apartment in building		
Available	104	100.0
Extent of communication with neighbours in multi-flat buildings		
Extensive communication	13	12.5
No communication	90	86.5
Don't know	1	1.0
Reasons for non-communication		
Not satisfied with neighbours	28	26.9
No common interests	64	61.5
Others	12	11.5
Willingness to share renovation costs with neighbours having lower income and not able to pay for renovation of their apartments in multi-flat buildings in case of increase of energy renovation costs by		
0%	92	88.5
Below 5%	8	7.7
6–10%	2	1.9
21–30%	2	1.9
Acceptance of covering energy renovation costs by other neighbours		
Accept	104	100.0

Source: Created by authors based on Streimikiene and Balezentis (2020).

The housing conditions of the respondents are provided in Table 2.7.

As one can notice from Table 2.7, households consisting of two, three or four members of households accounted for a similar share; about 30% of all respondents took part in the survey. One-member households accounted for 5% of all respondents. About 90% of the respondents lived in multi-apartment buildings. About 90% of the buildings were built during 1961–1993 period. On the basis of the survey, 90% of the respondents lived in three- or four-room apartments; 40% of the respondents lived in apartments of 20–50 m² while 40% of the respondents lived in apartments of 51–70 m². Above 50% of the respondents lived in five-floor multi-apartment buildings.

Above 70% of the respondents owned their apartments, while 40% of the respondents lived in multi-apartment buildings where housing associations were already established. This indicator suggests a high administrative burden of energy renovation in multi-flat buildings inhabited by survey respondents. In addition, more than 90% of the respondents did not know the energy class of their buildings. This shows the low awareness among respondents about the energy certification of buildings.

The main drivers of energy renovation in multi-flat apartments are provided in Table 2.8.

As can be seen from the data presented in Table 2.8, about 60% of the respondents during the cold season (heating season) pay from €21 to €50 per month and more than 30% of the respondents pay between €51 and €100 per month for electricity. At the same time during the warm season 51% of the respondents pay from €21 to €50 per month and 30% of the respondents pay between €51 and €100 per month. During the cold season 50% of the surveyed respondents pay between €101and €200 per month for heating and 33% of the respondents pay between €51 and €100 per month on average. During the warm season, 51% of the respondents pay below €20 per month for heating and 39% of the respondents pay between €21 and €50 per month for heating on average.

At the same time electricity costs do not change significantly during cold and warm seasons. More than 50% of the respondents indicated the sum to be between €21 and €50 per month for paying electricity bills on average during both the seasons. Heating costs increased from €21 to €50 per month in the warm season for 51% of the respondents and from €101 to €200 per month in the cold season for 50% of the respondents surveyed. Almost 60% of the households indicated the heating costs to be from €101 to €200 per month on average during the entire year. In order to estimate the changes in the costs and the ratio to the area of apartments, the data was grouped to evaluate the averages of these variables. The average area of apartments is 53.3 m² in this survey. During the cold season the total energy cost of one household is €187 per month on average. During the warm season the total energy costs of one household is €77 per month. The costs per m² are presented in Table 2.9.

As can be noticed from the information provided in Tables 2.8 and 2.9, energy costs, especially heating costs, are quite high in Lithuania taking into account the low income of households. Also, more than 90% of the respondents were unsatisfied with the heating comfort in their apartments. Also 100% of the respondents confirmed that the heating costs in their monthly bills did not correspond to low heating comfort in their apartments. Besides that, more than 90% of the respondents were thinking about energy renovation options, though they didn't implement this idea; 10% of the respondents replaced windows in their apartments and 26% of the respondents replaced exterior doors. In 9% of the households the roof of the building was repaired. However, about 60% of the respondents did not implement any partial energy renovation measures in their homes. As can be noticed from Table 2.8, though all respondents had the information about the available state support for multi-apartment building renovation, almost 90% of the respondents stated that the state support did not provide sufficient initiatives to renovate their homes. In Table 2.10 the main barriers of renovation are provided based on the survey results.

The main hurdles of energy renovation identified by 55% of the respondents are linked with the absence of a household association and the reluctance by respondents to take a loan for energy renovation, due to the fact that 35% of the respondents can't perform energy renovation without taking a loan. Above 90% of the respondents are convinced by the fact that the availability of one responsible agency bearing the responsibility for all the energy renovation issues of multi-apartment buildings will have a significant impact on their decision to initiate energy renovation works in their homes.

The results of assessment of WTP for energy renovation are shown in Table 2.11.

As one can notice from Table 2.11, 46% of the respondents indicated WTP below €10 per month for energy renovation and 42% of the respondents indicated WTP between €11 and €50 per month. Meanwhile, for more than 60% of the respondents WTP for energy renovation was from €2001 to €3000. Besides that, for 10% of the respondents WTP for energy renovation was from €1001 to €2000. For 10% of the respondents WTP was in the range between €3001 and €4000. For 12% of the respondents WTP was in the range between €4001 and €5000.

27% of the respondents stated that the WTP for energy renovation, if the renovation costs are included in their monthly energy bills, is from 31% to 40% of the increased monthly energy bills and for 15% of the households WTP for energy renovation makes from 41% to 60% of the increased monthly energy bills. At the same time, for 20% of the respondents WTP for energy renovation makes from 51% to 60% of the increased energy bills and for 14% of the respondents WTP is in the range from 61% to 70% of the increased monthly energy bills.

Taking into account that the average energy bill for heating during the cold season makes about €155 per month (Table 2.9), the WTP for energy renovation of the surveyed households in Lithuania under such a financing scheme is quite high and makes about €55 per month.

It is also necessary to highlight that above 90% of the respondents would choose to pay for energy renovation based on their increased monthly energy bills.

In Table 2.12 the results on the willingness to share renovation costs with neighbours having lower income are given.

All respondents indicated that there are households who cannot afford to renovate their apartments by themselves, but just 10% of the respondents expressed the willingness to share energy renovation costs with their lower income neighbours, in this way speeding up the process of energy renovation of multi-flat buildings. At the same time, all surveyed households agreed to accept help from other households to cover part of their energy renovation costs.

Above 90% of the surveyed households stated that the availability of one responsible agency solving all the organizational issues of energy renovation would motivate them to make decisions on energy renovation of their homes in multi-apartment buildings. Also, almost 90% of the respondents stated that they do not communicate with their neighbours due to various reasons. The dissatisfaction with neighbourhood was one of the main reasons of non-communication between neighbours in multi-flat buildings as well as the lack of common interests.

Also, the relationship between demographic factors and WTP and decision to share the renovation costs were explored by applying Chi-square test, the Median test, the Mann-Whitney U test, the Kolmogorov-Smirnov test and the Kruskal-Wallis test. It was found that work and income have a relationship with the decision to share costs with neighbours having lower income. Therefore, the decision of sharing is on the basis of the working situation and income of respondents. The rest of the demographic variables have no relationship. Since the p-value >0.05 in all the cases, thus, it was found that there is no relationship among the demographic variables and WTP for the full renovation of the building in total immediately before starting the process of energy renovation. Moreover, the Median test, the Mann-Whitney U test, the Kolmogorov-Smirnov test and the Kruskal-Wallis test suggested that there is no statistical significance among the relationship of demographics and acceptance to pay for the full renovation of the building in total immediately before starting the process. However, the ownership and the size of the apartment have a relationship with WTP for the full renovation of the building in total before starting the process of energy renovation.

The results of the study provided the barriers of energy renovation in Lithuania. Hypotheses 1–4 were not rejected. The absence of a housing association and the inability to agree on the decision to renovate buildings among Lithuanian households were defined as the main barriers of energy renovation in multi-flat buildings. Besides that, more than 90% of the surveyed households were convinced that the availability of a responsible organization taking all responsibility for all organizational issues linked to renovation will influence their decisions to renovate houses. Furthermore, more than 30% of the surveyed households in Lithuania are reluctant to take a loan necessary for energy renovation of their apartments and more than 80% of the respondents reflect that the governmental support for energy renovation does not provide adequate motivations to renovate their homes. More than 90% of the respondents would prefer to pay for energy renovation by an increase in monthly energy bills instead of paying the entire sum of energy renovation in advance.

The performed cross tabulation showed a significant impact of income and gender on WTP in advance for energy renovation before starting the process of energy renovation. Other demographic variables have no relationship with WTP for energy renovation. Besides that, the higher income provides for higher WTP for energy renovation. Females tend to have higher WTP for energy renovation.

The ownership status of the apartment and the size of the apartment also have an impact on WTP in advance for energy renovation before starting the process of energy renovation. The owners of apartments and the owners of bigger size apartments have higher WTP for energy renovation.

The work and income both have an impact on the decision to share the cost of energy renovation with neighbours having lower income. Consequently, the decision of sharing the costs of energy renovation is linked with the working status and the income of surveyed households. Other demographic variables have no relationships with WTP for energy renovation. Higher income respondents are more willing to share the costs of renovation. Employees employed in the private sector tend to have a higher willingness to share renovation costs with low-income neighbours.

The other studies on WTP for energy renovation (Poortinga *et al.*, 2003; Jakob, 2006; Grosche & Vance, 2009) have examined the impact of social-economic characteristics of the respondents on implementation of energy efficiency measures in other European Union Member States. In these works, the heating system purchase decisions as a function of property and household characteristics were investigated as well. These studies identified that income, age, gender, education, ownership of apartment and environmental awareness were the main factors having an impact on WTP of households.

Though the current study didn't find the linkages between the age and education of respondents and their WTP for energy renovation, however, income, ownership, size of the apartment and gender were confirmed to be central drivers of WTP for energy renovation in Lithuania.

2.3 Conclusions

The assessment of WTP of renewable micro-generation technologies were performed among Lithuanian households by applying the discrete choice experiment approach. The results showed that Lithuanian households are willing to pay extra only for micro-generation technologies based on solar PV. Therefore, these technologies require less public support and high tariffs for solar energy are likely to reflect the full benefits of these technologies.

In the case of biomass boilers, solar thermal and micro-wind technologies, which are also subject to high-price premiums tariffs, energy consumers indicated a negative WTP and showed that they need additional compensation in order to be able to choose one of these technologies. Again, negative WTP values should be used for guidance purposes only and not strictly understood as a need for additional financial support.

The differences in WTP of the four renewable micro-generation technologies have some impact on Lithuanian energy policy. In fact, in addition to premium rates, the country applies various support measures for RES like financial support from the EU Structural Funds and other funds, pollution tax incentives for stationary pollution sources (for burning biomass), etc.

Positive WTP for solar PV shows that these technologies are likely to dominate the Lithuanian market. This is a good option for most urban housing, but some improvements may be needed in rural areas. More specifically, rural areas can use crop and forest residues to install biomass boilers. In addition, micro-wind generators may also be an option for rural areas with abundant land area. Therefore, support for RES should be differentiated in rural and urban areas, depending on household preferences and assessment of WTP for micro-generation technologies.

Although all micro-generation technologies analysed have a zero-emission standard according to the principles established by the International Commission on Climate Change, biomass technologies are associated with some environmental impacts related to atmospheric emissions. The possibility of sharing micro-generation technologies turned out to be insignificant for Lithuanian households. This can be explained by certain factors related to the Lithuanian context as a bad experience of collectivization during soviet times.

Further research is necessary to investigate this mode deeply. First, the assessment may include attitudes towards the environment and other socio-economic characteristics. Second, the analysis should be repeated for different customers groups and at different time periods. Third, the setting of preferences in surveys can involve a number of issues such as risk-taking, social attitudes, time discounting and changing preferences. These and other issues related to cognitive limitations should be taken into account in future assessments of WTP for renewable micro-generation technologies.

The additional attributes (e.g. ease of maintenance, user-friendliness, system flexibility) could be included in the model. An expert survey can be conducted to prioritize alternative technologies based on social needs that individual consumers may ignore. Additional micro-generation technologies suitable for multi-apartment buildings may be taken into account when expanding the scope of the analysis.

The study of WTP for energy renovation in multi-apartment buildings was performed in Lithuania by applying the contingent valuation method. Though other studies of WTP for energy renovation of multi-flat buildings confirmed that income, age, gender, education, environmental awareness and ownership of an apartment are important drivers of household WTP for energy renovation, this work did not find a linkage between the age and education of the respondents and their WTP for energy renovation, but income, ownership, apartment size and gender were found as significant drivers of households WTP for energy renovation in Lithuania. Besides that, the performed study on WTP for energy renovation in Lithuania showed that the absence of a housing association and problems linked with agreement between apartment owners on energy renovation of multi-flat buildings are the most important organizational barriers to deep energy retrofitting in Lithuanian multi-apartment buildings.

Also, the study provided that more than 90% of households believe that the person or organization responsible for all organizational issues related to the energy renovation of a multi-apartment building would be an important driver for their decision making on energy renovation of their homes. However, the study found that state support is not adequate to encourage energy renovation of multi-apartment buildings as more than 80% of the surveyed households stated that government support for energy renovations is not sufficient for them. Also, above 30% of the respondents were reluctant to take a loan for energy renovation of their homes.

The survey provided that above 90% of the surveyed households did not want to share the energy renovation costs of low-income neighbours and in such a way to smooth and speed up the energy renovation process of their apartments in the same building. This is related to the situation in Lithuania as households have in general low income and are not able to bear the costs of energy renovation of other even more vulnerable neighbours.

The study found that WTP for energy renovation of multi-flat buildings in Lithuania makes about €55 per month and more than 90% of the surveyed households prefer to pay for energy renovation by including renovation costs in their monthly energy bills. Just less than 10% of the respondents of performed survey stated that they want to pay for the full costs of energy renovation in advance.

The main policy implications of the performed study on WTP in Lithuania is the recommendation for the application of Energy Service Company Obligation (ESCO) model for energy renovation of multi-flat buildings in Lithuania. This ESCO model will allow to include the costs of energy renovation in monthly energy bills and to take all responsibilities for organizations of energy renovation of multi-flat building, covering the negotiation with households and reaching the common decision on energy renovation of multi-apartment buildings. In this way the ESCO model allows to overcome the two main energy renovation barriers of multi-flat buildings in Lithuania: lack of capital and reluctance to take a loan as well as an inability to reach a common agreement on energy renovation of multi-apartment buildings among the owners of apartments.

3

Lithuania's Climate Change Mitigation Policy Related to Energy Consumption in Households

3.1 Climate Change Mitigation Policy Priorities

The European Union's (EU) Climate and Energy Policy Strategy is the main strategic document for climate mitigation in the EU, setting priorities and key orientations. On 22 January 2014, the European Commission presented the EU 2030 climate and energy framework for 2020–2030. The aim of this document is to start identifying policies for these priority areas after 2020 when the current strategy expires. The strategy sets out further measures to mitigate climate change, with the aim of reducing greenhouse gas (GHG) emissions by 80–95% by 2050, as compared to the levels of 1990. It also aims to reduce the EU's dependence on energy imports, often from politically unstable regions and modify and upgrade energy infrastructure and provide a stable regulatory framework for potential investors, while achieving the 2030 GHG reduction target.

The main obligations are to further reduce GHG emissions by setting a target of reducing the amount by 40% below 1990 levels by 2030. At the same time, the aim is to achieve a renewable energy share of at least 27% of the final energy consumption by giving the Member States the freedom to set national targets. The aim is also to improve energy efficiency through future amendments to the Energy Efficiency Directive. The strategy foresees the reform of the EU emissions trading system (ETS) and the creation of a market stability reserve.

In implementing the strategic goal of climate change mitigation, the National Strategy for the Lithuanian Climate Change Management Policy (Ministry of Environment of Republic of Lithuania, 2012) provides for a contribution to the EU short-term climate change mitigation goals by 2030 and an internal EU effort to reduce GHG emissions by at least 40% by 2030, as compared to 1990 and ensure that GHG emissions in the sectors participating in the EU ETS are reduced, respectively, by 43% and by at least 30% in non-ETS sectors, as compared to 2005.

In order to ensure the implementation of Lithuania's short-term climate change mitigation goals by 2030, it is necessary to reduce GHG emissions in the EU ETS sectors by 43% as compared to 2005 and reduce GHG emissions in non-ETS sectors by 9% as compared to 2005. It is necessary to ensure at least 0.9% of the country's GDP in 2030 to be allocated to achieve short-term climate change mitigation goals. Also, it is necessary to achieve the implementation of Lithuania's main indicative medium-term and long-term climate change mitigation goals: in the medium term to reduce GHG emissions by 70% by 2040 compared to 1990. Also it is necessary to reduce GHG emissions in the long term by 2050 by 80%, compared to 1990, and by 20% to cover sinks from the land use, land use change and forest (LULUCF) sector by applying environmentally safe technologies for geological storage and carbon capture and use of carbon dioxide. The long-term goal is to reduce GHG emissions by 80% by 2050 compared to the levels in 1990 and to cover 20% with sinks from the LUCF sector and by applying environmentally safe technologies for the geological storage and use of carbon dioxide.

3.2   Policies to Promote Renewable Energy Sources

According to the data of 2019 in Lithuania, the share of renewable energy sources (RES) in the total final energy consumption accounted for 24.21%. These results were mainly determined by the share of RES in the heating sector, which amounted to 46.50%. The share of RES in electricity generation accounted for 18.41%, while in the transport sector the share of RES was 4.33%.

The RES development in Lithuania is carried out by taking into account EU and national strategic documents and legal acts. The main RES development policies and measures are established in the updated National Energy Independence Strategy (NEIS) (Ministry of Energy of the Republic of Lithuania, 2018), the National Strategy for Climate Change Management Policy (the Ministry of Environment of the Republic of Lithuania, 2012) and the Law of the Republic of Lithuania on Energy on Renewable Sources and integrated into the National Energy and Climate Action Plan of the Republic of Lithuania 2021–2030 (the Ministry of Environment of the Republic of Lithuania, 2020). The following Lithuania's RES development priorities and specific RES development goals set out in Lithuania's strategic documents are discussed below.

3.2.1   Priorities for Promoting Renewable Energy Sources

Lithuania reached the EU goal of 23% for 2020 and exceeded it in 2014 when the share of RES in the total final energy consumption amounted to 23.66%.

As a result, Lithuania transferred part of the surplus in 2017 to Luxembourg and became the first EU Member State to sign a cooperation agreement on the transfer of quotas from statistical RES. Lithuania plans to reach the goal of 45% of RES in gross final energy consumption by 2030, and 80% by 2050.

The share of RES in the final energy consumption is ensured by increasing the share of RES in the electricity, transport and heating and cooling sectors.

Table 3.1 shows the renewable energy targets for final energy consumption, electricity, transport and heating and cooling.

As it was clear from situation in 2019, that the goal to reach 30% of RES in the electricity sector by 2020 set out in the NEIS of Lithuania will not be achieved; therefore, the commitments of Lithuania in this sector (RES-E) for 2050 are set based on the RES usage situation of 2019. The integration of RES into the transport sector is inefficient and too slow, and the process is expected to accelerate with the adoption of the law on alternative fuels. The integration of RES into the transport sector is inefficient and too slow, and it is expected to speed up the process with the adoption of the Alternative Fuels Act.

Since 2015, after the allocation of the full support quota set in the Law on Energy from Renewable Sources until 2020, the provision of support to RES-E has been suspended; therefore, the development of RES-E has slightly slowed down. In 2018, the share of RES-E in the final energy consumption amounted to 18.41%; new promotional quota allocation auctions started at the end of 2019 to promote the development of RES-E; therefore, the results

TABLE 3.1

Objectives of promotion of renewable energy sources as established in strategic documents

Obligations of Lithuania	2017	2018	2020	2022	2025	2027	2030	2050
Share of renewable energy sources in final energy consumption, %	26.04	2.21	30	32.70	36.45	39.75	45	80
Share of renewable energy sources in electricity production, %	18.25	18.41	30	25.55	31.48	36.70	45	100
Share of renewable energy sources compared to the final energy consumption in the transport sector	4.29	4.33	10	6.69	9.23	11.46	15	50
Share of renewable energy sources in heating and cooling production, %	46.50	47.30	—	53.9	63.1	66.9	67.2	—
Share of renewable energy sources in district heating supply, %	68.7	67.5	70	78.8	89.3	91.3	90	100

Source: Created by authors based on the Ministry of Environment of the Republic of Lithuania (2020).

TABLE 3.2

Development of the use of renewable energy sources in the electricity sector

Indicator	2020	2022	2025	2027	2030
Final electricity consumption, ktoe	897	863	850	844	878
Hydroelectric power plants, ktoe	42.6	42.6	42.6	42.6	42.6
Wind power plants, ktoe	98.4	98.4	247.1	307.3	382.5
Solar power plants, ktoe	5.9	7.4	36.6	62.4	74.0
Biofuel power plants, ktoe	25.1	47.5	50.1	50.1	50.1
Biogas power plants, ktoe	10.9	10.9	10.9	10.9	10.9
Industrial and municipal waste cogeneration plant	7.2	20.7	20.7	20.7	20.7
RES-E, ktoe	190.1	227.5	384.8	445.0	520.2

Source: Created by authors based on the Ministry of Environment of the Republic of Lithuania (2020).

are likely to be visible only by 2023 when the winners will build the power plants and start generating electricity. The main measures used in Lithuania until 2020, which promote the development of RES-E, are the financial promotion of consumers, which will have a minor impact on the development of RES-E; therefore, it is likely that in 2020 the 30% target set in the NEIS will not be achieved.

Wind energy is estimated to remain the main source of electricity generation with at least 70%, solar energy 3%, biofuels 9%, hydropower 18% (Table 3.2).

The development of RES has been taking place in Lithuania since 2002, when the first hydropower plants and wind power plants were built. Given that the development of hydropower plants in Lithuania is limited by environmental laws, it is estimated that the development of these power plants will not take place during the period 2020–2030. The total installed capacity of new wind power plants will amount to 1,322 MW during 2020–2030. The development of biogas and biofuel power plants is not foreseen for the 2020–2030 period.

One of the most important directions in the field of promotion of RES is the promotion of active electricity consumers able to consume their generated electricity for own needs. Such active consumers must receive a remuneration appropriate to market conditions for the surplus energy supplied to the grid. Lithuania also aims to promote the active participation of local communities in investing in RES facilities managed by joint ownership.

The renewed NEIS includes ambitious goals for the development of active electricity consumers:

- 2% compared to the total number of consumers by 2020;
- 30% compared to the total number of consumers by 2030

In the transport sector, the aim is to gradually shift towards less fuel and electricity consumption; therefore, to implement the commitments to the

EU, it is aimed to achieve a 10% share of RES in 2020 and 15% by 2030 (Table 1.1). However, Lithuania, like other Member States, has difficulties in implementing the RES-T part by 2020 due to the relatively large investments in the renewal of the vehicle fleet, most of which account for almost 1.5 million passenger cars, of which 69% are diesel cars, and their exploitation age is 15 years.

For this reason, it is likely that the 2020 target will not be reached by 2020 and the RES-T share will be around 5%.

More efficient actions are currently being taken to promote the growth of the RES-T share, with higher standards for a mandatory blending of biofuels coming into force on 1 January 2020, setting out key principles for promoting the use of alternative fuels and the use of less polluting vehicles. It is planned to focus on the use of second generation advanced liquid and gaseous biofuels and electrification of the vehicle fleet and railway system. The law also aims at establishing a clear long-term perspective for market participants in the transport sector and providing for transitional alternatives.

The main goal of Lithuania in the heat sector is a consistent and balanced renewal (optimization) of district heating systems, ensuring efficient heat consumption, reliable and economically attractive (competitive) supply and production, enabling the introduction of modern and environmentally friendly technologies, using local and renewable energy resources, ensuring system flexibility and favourable investment environment. Following the best practice of the EU countries, the transition to the fourth generation (4G) district heating, integrating solar power plants into district heating networks and promoting the use of surplus and waste heat for heating buildings must be encouraged in Lithuania.

In Lithuania, the district heating system is an integral part of the overall energy sector and is closely related by technological and energy flows to the electricity system, fuel supply and other systems. Well-developed district heating systems operate in all cities of Lithuania, of which about 53% of all buildings are supplied in the country and about 76% of all buildings are supplied in cities. The main consumers of district heating services are residents living in multi-apartment buildings.

The total installed capacity of heat production facilities in district heating systems is 9,582 MW. Lithuania will aim for district heating produced from renewable and local energy sources to account for 90% of the total district heating by 2030. It is forecasted that 647 ktoe of fuel will be consumed in private households for heat energy production in 2020 and 643 ktoe of fuel by 2030.

The efficiency of heat and hot water production technologies in the decentralized sector is rather low and there is considerable potential for energy savings. The conversion of primary energy sources is also possible in this sector, which can significantly improve the heat supply conditions of the population and encourage more efficient use of RES, some of which could be used in other sectors.

TABLE 3.3

Development of the use of renewable energy sources in the heating and cooling sector, ktoe

Development of the use of renewable energy sources	2020	2022	2025	2027	2030
Energy demand for decentralized heat production	1,303.0	1,337.1	1,506.0	1,535.1	1,506.0
Decentralized heat production from RES	647.2	630.0	627.9	645.3	643.0
Coal for decentralized heat production	183.5	179.3	138.4	95.8	88.8
Petroleum products for decentralized heat production	144.5	143.5	138.0	130.9	129.0
Use of natural gas for decentralized heat production	554.5	539.8	522.2	510.6	495.8
Energy demand for district heating	1,072.8	1,028.0	952.1	916.8	876.6
District heating from RES	655.8	707.1	878.0	889.8	863.0
District heating, including waste heat	904.1	868.3	828.8	709.9	764.2
District heating for own use	9.2	8.9	8.5	8.2	7.8
Losses of district heating on routes	117.6	113.0	108.0	104.1	99.4
Percentage of energy used in total heating from renewable sources energy	50.9	53.9	63.5	66.9	67.6
Percentage of energy from renewable sources used for district heating	71.5	78.7	86.9	91.3	90.9

Source: Created by authors based on the Ministry of Environment of the Republic of Lithuania (2020).

The total share of RES in the heating and cooling sector will reach about 50% by 2020, where the major part will be heat produced from local biofuels. Additional policy measures (such as the installation of solar and heat pumps, low-temperature heating, use of waste heat), increased energy efficiency and further centralization in decentralized heat production are expected to reduce the demand for all fuels. Due to the specificity of the building stock, the energy demand for the cooling sector in Lithuania is negligible (Table 3.3).

Lithuania's goals in the heating and cooling sector are ambitious, however closely related to energy efficiency: in the field of decentralized heating and district heating, energy demand will decrease by 2030. New technologies (heat pumps, modern biofuel boilers, etc.) and renovation of multi-apartment buildings will have the largest power supply. If the additional measures envisaged are implemented, a higher share of RES in heating and cooling than currently foreseen can be achieved.

In Lithuania, the RES development in the electricity, transport and heating sectors is promoted through financial (state budget appropriations, Climate Change Programme funds, EU support funds, revenue from the implementation of agreements on statistical energy transfers or joint projects, tax incentives) and non-financial measures (obligations, information and regulatory measures).

The basic principles of promotion of the use of RES are as follows:

- The gradual integration of RES into the market: the most cost-effective technologies must be developed, technology maturity must be taken into account and trends in their near future progress should be considered.
- Affordability and transparency: the design of the scheme to promote RES must be based on the market principle, distort it as little as possible and ensure the minimum financial burden for energy consumers, clarity and non-discriminatory competitive environment.
- Active participation of energy consumers: the increasing share of RES in relation to the total balance of energy sources must encourage the decentralized generation of electricity, provide consumers with the possibility to use the energy produced from RES for their own use and receive remuneration for surplus energy supplied to the grid in accordance with market conditions, as well as consumer behaviour and energy demand and supply management solutions.

In accordance with the Directive (EU) 2018/2001 on the promotion of the use of energy from renewable sources (RED II), Lithuania assessed the need to build new infrastructure for district heating and cooling produced from RES in order to meet the national target. Under this Directive, in order to promote the use of energy from renewable sources in the heating and cooling sector, each Member State shall endeavour to increase the percentage of energy from renewable sources in that sector by an indicative average of 1.3 percentage point in that sector as an annual average for the 2021–2025 and 2026–2030 periods, starting from 2020 as a percentage of renewable energy in the heating and cooling sector, expressed as a national percentage point of the final energy consumption and calculated in accordance with the methodology set out in Article 7, without prejudice to Article 23(2) of the Directive. That increase shall be limited to a reference 1.1 percentage point for those Member States which do not use waste heat and cooling. The Member States shall, where appropriate, give priority to the best available technologies. Each Member State, when calculating its percentage of energy from renewable sources in the heating and cooling sector and its annual average increase according to that share (Ministry of Energy of Republic of Lithuania, 2014):

- may include waste heat and cooling up to 40% of the average annual increase;
- where its percentage of energy from renewable sources in the heating and cooling sector is more than 60%, it may consider that this percentage fulfils the requirement of an average annual increase;
- where its percentage of renewable energy in the heating and cooling sector is more than 50% but less than 60%, it may consider that this percentage fulfils the requirement of half the average annual increase.

The centralized cooling network in Lithuania is not developed. Residential and commercial premises are cooled individually, using electricity for a cooling generation. The annual reference demand for cooling in Lithuania ranges from 5 to 6 TWh. The need for cooling was determined on the assumption that Lithuania's demand for cooling is ~60 kWh/m² per year, taking into account climatic conditions; however, in order to develop this sector, it is necessary to consider that it would be useful to do this only when those buildings already equipped with a centralized (common mechanical) ventilation system, i.e. offices, supermarkets and new high-energy class multi-apartment buildings, are connected to the grid, as an investment in old multi-apartment buildings would not be justified by district cooling. In this case, the annual cooling demand would be reduced to 2–3 TWh. In the perspective of 2021–2030, the priority axes for improving regulatory market conditions in the heat sector are as follows:

- Establishment of a regulatory environment promoting the attraction of investments and providing a non-discriminatory environment for all participants in the district heating market;
- Increasing transparency in the biofuel market;
- Promotion of the supply of heat produced by district heating in buildings and prioritizing urbanized areas in order to reduce air pollution;
- Assessment of the current situation and forward-looking development of heat supply in the decentralized sector, foreseeing rational development directions, estimating changes in heat production technologies and increasing efficiency of heat production and consumption;
- Assessment of the current situation in the cooling sector, carrying out a prospective analysis, and establishment of guidelines for the most rational solutions of self-sufficiency in cooling.

The technical tasks of implementing the solutions are as follows:

- Assessment of the possibilities of the use of waste heat collection and use, solar light and heat energy technologies, heat pumps, low-temperature heating and heat storage facilities for the production of district heat and, where this is economically justified, their implementation;
- Implementation of a remote heat meter reading system;
- The rational development of high-efficiency cogeneration plants which increase the potential for local electricity production;
- Existing biofuel combustion installations, existing heat transmission installations and their systems as well as heat points of buildings and/ or heating and hot water systems must be modernized in due time by creating technical conditions, where this is economically justified, for individual heat energy consumption regulation for each consumer.

3.2.2   Promotion of the Use of Renewable Energy Sources in Households

Renewable energy technologies are gradually taking the road to the renovation of multi-apartment buildings in Lithuania; however, there are many problems preventing house administrators from implementing the latest technologies enabling residents of multi-apartment buildings to live in a more economical way. In Lithuania, there are multi-apartment buildings that already have solar collectors for hot water preparation. This is particularly relevant for those multi-apartment buildings that have significant losses in the hot water circulation line. This solution allows residents to become independent of the hot water supplier and pay off within seven years. Thermostatic regulators are also being introduced, which allows residents of multi-apartment buildings to save between 10% and 25% of heat energy. Energy-saving lighting systems are installed in the stairways of apartment buildings managed by the housing administrator, energy consumption data of multi-apartment buildings are constantly monitored, the programme "5 Steps" ("5 žingsnių") is implemented, energy consumption decreases by about 20% without any renovation, but simply by automating the heat point, implementing real-time controlled heat systems and carrying out minimum works to reduce heat loss in co-apartments (ESTEP, 2019a).

Thus, the renovation of multi-apartment buildings currently being carried out in Lithuania is very inflexible, thoughtless, the system is constantly changing and cumbersome, and the financing is not transparent, which hinders the process of house modernization. There are a number of measures that do not always increase the energy efficiency of the buildings after renovation, and residents do not always see the promised effect. Although the energy efficiency of the building does not increase or increase to a minimum, this renovation still receives 30–40% of support. It is recommended to allocate money for the energy class of multi-apartment buildings.

The financial support for prosumers was approved in 2018 and is planned to continue until 2030. The activities supported are the installation of small-scale solar power plants. It is planned that as of 2024, installations with an installed capacity of 696 MW will be built with this support. In 2015, a scheme for electricity prosumers was created to promote the active participation of electricity consumers in the market. By 2030, the aim is to reach 30% of prosumers, compared to the total number of electricity consumers. In order to ensure that all electricity consumers can benefit from the prosumer scheme, funds from the EU Structural Funds and the National Climate Change Programme will be allocated for the purchase of the power plant. Since 2019, €323 per 1 kW of consumer support has been granted. In total, more than €16 million of EU funds are planned to be invested by 2023, with four calls planned during this period.

Additional measures include:

- RES-E exemption from excise duty. Electricity produced using RES will be exempted from the obligation to pay excise duties. This provision will apply both to electricity produced in Lithuania and imported in Lithuania.
- Guarantees of RES-E origin. Origin guarantees are issued to RES-E producers. Guarantees of origin are also issued to RES-E producers who have won the auction and receive the price premium for electricity.
- Sale and purchase agreements of RES-E. RES-E producers are granted the right to sell electricity to final consumers under agreements for the sale and purchase of electricity from renewable sources without holding a licence of an independent electricity supplier. Such producers will still have to meet the requirements of an independent electricity supplier.
- Electricity prosumers. Natural and legal persons planning to produce electricity in the power plants using solar, wind and biomass resources, whose installed capacity may not exceed 500 kW, may become prosumers.

The prosumer is given the opportunity to "accumulate" the electricity not consumed by them for own needs and for the needs of the economy in the electricity grids from 1 April of the current year until 31 March of the following year. The producer shall pay a fee on the use of electricity grids for the amount of electricity "accumulated" and recovered from the electricity grids. The amount of electricity supplied to electricity grids in excess of the amount of electricity consumed by the prosumer during the storage period will not be transferred to the next storage period.

Prosumers can install power plants themselves or buy them through bilateral contracts from third parties, thus creating the possibility to become a prosumer for persons living in multi-apartment buildings. The power plant of the prosumer can also be remote from the place of consumption of electricity. In this case, the power plant must be owned or controlled by the prosumer.

It is foreseen that the prosumer scheme, allowing electricity not consumed for own use and for the economy to be "build" in electricity grids, will continue until 1 April 2040. The assessment of the conditions will be carried out in accordance with Directive 2018/2001.

In order to promote the development of prosumers, Prosumer Alliance is being developed, with representatives of RES business associations and consumer organizations. The aim of the Alliance is to promote small green energy for the population and business by promoting consumer education and awareness of green energy and the opportunities to have their own power plant and self-generated electricity. The aim is also for all partners of the Alliance to offer high-quality and competitive products and services.

Another goal is to intensify cooperation in search of innovative solutions that would help to accelerate the development of small green energy in Lithuania.

In 2019, the Ministry of Energy conducted a public consultation to identify the main obstacles and drivers for the establishment of renewable energy communities. The main obstacles to the establishment of renewable energy communities were identified as follows:

- traditionally sluggish organizing of the residents of the country's cities and regions into communities;
- it is difficult to create an association for multi-apartment buildings whose apartment owners in the associations uniting them do not take decisions unanimously;
- obtaining a bank loan;
- protected areas;
- complex legal requirements;
- often changing legal framework;
- lack of space for the construction of power plants in major cities.

The following incentive actions have also been identified:

- closer cooperation of the distribution system operator;
- application of incentives;
- leadership of public administration institutions (consultations, practical examples);
- assistance and coordination of public institutions.

In the light of the observations made during this public consultation, the Law on Energy from Renewable Sources was amended in 2019 and a scheme for renewable energy communities was established. A renewable energy community shall be defined as a legal status granted to a public establishment which meets the established criteria and which, within the defined territory, manages and develops by the right of ownership installations for the production of energy from renewable sources and has the right to produce energy, consume it, store energy in storage facilities and sell it (Ministry of Environment of the Republic of Lithuania, 2020).

Natural persons, small- or medium-sized business entities and/or municipalities, may be participants of a renewable energy community and from whom:

1. at least 51% of the members having a voting right or shareholders are natural persons whose place of residence has been declared in a municipality wherein an energy production facility (equipment) is planned to

be built or installed, or in other municipal elderships bordering to this municipality;

2. each member or shareholder holds not more than 20% of the voting rights of another energy undertaking.

Renewable energy communities are granted the right to participate in the auctions without undertaking to produce and deliver to grids the total amount of electricity awarded at the auction. They may also carry out any activity in the energy sector with an appropriate authorization. With a view to facilitating the establishment of associations of energy from renewable sources, municipalities are obligated to assess and publish the sites at which installations for energy production from a RES may be constructed or installed. The possibilities for facilitating the transmission of electricity from renewable sources within the EU are currently being assessed.

3.3   Energy Efficiency Policies and Renovation of Multi-Apartment Buildings

Improving energy efficiency and the use of RES are the most important measures to reduce GHG emissions in the energy sector. The National Strategy for Climate Change Management Policy of Lithuania aims to ensure that the implemented energy efficiency improvement measures (renovation of residential and public buildings, zero-energy buildings, high-efficiency heating and cooling systems, development of clean electrified various modes of transport, high-energy class industrial and household appliances, etc.) contribute to achieving a decrease in primary and final energy intensity by two times compared to that in 2017 (the Ministry of Environment of the Republic of Lithuania, 2012).

If the use of RES requires significant investment and public support for integrating external benefits related to the use of RES, increasing energy efficiency is the most effective climate change mitigation measure in the energy sector, which allows significant savings in energy bills. However, even in such a situation, the potential for increasing energy efficiency is not fully realized, as there are various barriers and market failures that hamper the implementation of energy efficiency improvement measures. As a result, many governments are committed and are implementing a wide range of energy efficiency and energy-saving measures. Lithuania is no exception. The following sections analyse the priorities and key strategic documents of Lithuania's energy efficiency improvement policy as well as their objectives. Since in Lithuania the largest potential for energy efficiency improvement is found in buildings, and especially in multi-apartment buildings, the programme for the renovation of multi-apartment buildings is given the most important attention; therefore, when reviewing its objectives and results, the main shortcomings and problems related to the realization of the potential for increasing energy efficiency in multi-apartment buildings are mentioned. At the same time, as the main problems are identified, other sections of the monograph seek for an answer of how to address these shortcomings and effectively apply the methods of preferences expressed to mainstream public opinion in public policy decisions in this area.

3.3.1 Priorities of Energy Efficiency Policies

Increasing energy efficiency is one of the most important goals of the EU and national goals of Lithuania. The main EU legal instrument for the implementation of energy efficiency policies is Directive 2012/27/EU of the European Parliament and the Council on energy efficiency, amending Directives 2009/125/EC and 2010/30/EU and repealing Directives 2004/8/EC and 2006/32/EC (EC, 2012).

On 25 October 2012, to implement Directive 2012/27/EU on energy efficiency (EC, 2012) the National Energy Efficiency Action Plan in Lithuania was drafted and approved (Ministry of Energy of Republic of Lithuania, 2014). Every three years, by 30 April, an updated energy efficiency action plan is submitted to the European Commission by providing relevant information on Lithuania's energy efficiency target, energy efficiency targets for the whole economy or specific sectors, a description of the adopted efficiency policy measures, the results of which include final energy savings in economic sectors and the amount of final energy savings expected by 2020. It is also important to provide data on other efficiency policy measures (e.g. GHG reductions, improved air quality, place of employment creation) and the implementation of the efficiency policy measure budget. If no data are available on energy savings for each efficiency policy measure, the reduction in energy consumption at the sector level due to the efficiency policy measures taken (or combination of them) is reported. The first and second efficiency plans must provide the results related to the specified energy-saving target of the Republic of Lithuania; the methodologies used to calculate energy savings, lists of public authorities and other undertakings that have drawn up energy efficiency improvement programmes; a description of the energy-saving measures applicable to electricity, heat or gas suppliers; the heat values of fuel combustion; the total number of energy audits carried out over the previous three years; and progress in the development of cogeneration. It will also include measures to improve the cost-effectiveness of the energy consumption of network infrastructure, measures taken to ensure and develop energy demand management; efficiency policy measures to remove regulatory and non-regulatory barriers to energy efficiency; a description of the energy audit and energy use system; an updated forecast of expected gross primary energy consumption in 2020 and expected primary energy consumption in industry, transport, households and services sectors (ESTEP, 2019b).

The indicative energy efficiency target of the Republic of Lithuania is based on the final energy consumption and final energy savings. This target will be expressed at the absolute level of primary energy consumption and the final energy consumption for year 2020 and will be explained how it is calculated.

The main strategic documents of Lithuania pay great attention to the improvement of energy efficiency and energy savings in residential buildings during the renovation of multi-apartment buildings. The objectives of the public policy field of energy efficiency and housing renovation set in Lithuania's progress strategy "Lithuania 2030" in the field of "Smart Economics" include favourable conditions for business, territorial cohesion and ecological sustainability (the Ministry of Environment of Republic of Lithuania, 2012). The National Progress Programme 2014–2020 (the Government of the Republic of Lithuania, 2012) which implements the Strategy defines the priority "Environment favourable for economic growth" setting the goals relevant for energy efficiency and housing renovation: to promote the sustainable use of resources and ensure the stability of ecosystems.

The objectives set out in the Programme Implementation Plan of the seventeenth Government, adopted in 2017, are in line with the objectives of promoting the sustainable use of resources and ensuring the stability of ecosystems as set out in the National Progress Programme 2014–2020.

In addition to the general documents set out in the country-wide strategic objectives, energy efficiency and housing renovation objectives are also set in sectoral medium-term strategies. The Energy Efficiency Action Plan for 2017–2019 had set the overall target of 11.674 TWh of final energy savings by 2020, calculated on an aggregated basis. This is envisaged through seven energy efficiency improvement policies: taxes and excise duties on fuel; renovation of multi-apartment buildings; improving the energy efficiency of public buildings; energy audits in the industry; agreements with energy suppliers on consumer education and advice; energy savings agreements with energy companies; and replacement of boilers in households.

The interinstitutional action plan for the implementation of the goals and objectives of the National Strategy for Climate Change Management Policy for 2013–2020 set the following goals (Ministry of Environment of Republic of Lithuania, 2012,:

1. to seek that GHG emissions in the sectors participating in the EU Emissions Trading Scheme in 2020 do not exceed 8.53 million tonnes of CO_2e;
2. to seek the integration of climate change adaptation goals, objectives and measures into the most sensitive sectors of the country's economy;
3. to integrate climate change adaptation and climate change mitigation goals, objectives and measures into the strategies and plans of the

country's economic sector (energy, industry, residential development, agriculture, transport);

4. to contribute to international cooperation on climate change.

The strategic goal of Lithuania's climate change mitigation policy is to achieve a much faster growth of the country's economy than the growth of GHG emissions. This will be done by increasing energy efficiency and the use of renewable energy.

In summarizing the objectives set, it appears that in the area of energy efficiency and housing renovation, the aim is to reduce not only GHG intensity but also energy intensity, i.e. energy per unit of GDP. All of this is aimed at ensuring the country's economic growth, international competitiveness and maximizing the benefits of the country's development of energy efficiency measures.

Energy efficiency is usually measured in terms of primary and final energy intensity, which indicates the amount of energy costs incurred in developing a specific quantity of goods and services in the country (relative to GDP of the economy's energy consumption). Industry, buildings and transport have the greatest potential for improving energy efficiency, given the feasibility of efficiency measures. In industry, energy consumption at product cost remains high and is 20% higher than the EU average and requires more efficient and modern technologies and energy management tools to reduce energy costs and improve undertakings' competitiveness. The transport sector accounts for around 38% of the total final energy, which makes it necessary to increase energy efficiency and implement related energy efficiency measures in this sector.

Multi-apartment residential buildings use the largest amount of heat energy, i.e. 54% of final heat consumption. It is in this field, which accounts for 60% of all Lithuanian buildings by area, the biggest potential for saving the amount of heat energy is observed.

Implementing the objective of increasing energy consumption efficiency as specified in the NEIS, the following energy-saving objectives are identified:

- by 2020, to ensure the implementation of EU requirements for Lithuania in the field of energy efficiency improvement, i.e. by calculating the total energy savings of 11.67 TWh and financial compliance with these requirements;
- by 2030 to ensure that the primary and final energy intensity is 1.5 times lower than in 2017;
- by 2050, to ensure that primary and final energy intensity is about 2.4 times lower than in 2017.

The main directions for achieving the energy efficiency improvement goals in Lithuania are as follows:

- to promote the integrated renovation of multi-apartment residential and public buildings (prioritizing renovation of residential blocks) by 2020 and in multi-apartment residential and public buildings (totalling annual energy savings) to save between 2.6–3 TWh and 5–6 TWh by 2030;
- to implement the rapid development of low-energy and energy-efficient industries, the deployment and acquisition of the latest and environmentally friendly technologies and equipment;
- to increase energy efficiency in the transport sector by upgrading the vehicle fleet, moving towards modern and efficient public transport, optimizing transport and alternative fuel infrastructure, electrifying it or using alternative fuels.

Increasing energy efficiency in Lithuania is carried out in accordance with the following principles:

- economic feasibility: priority should be given to the most cost-effective energy efficiency improvement measures in the context of energy efficiency improvement objectives;
- active education and training of energy consumers: as energy consumers can contribute to energy efficiency objectives by changing their behaviour and habits, the training and education of energy consumers must be strengthened;
- competition: by enabling investors to increase energy efficiency to compete with each other for the implementation of the most economically beneficial projects, fulfilling energy efficiency commitments or competing for state incentives.

3.3.2 Multi-Apartment Building Renovation/Modernization Programme

In order to implement the requirements of Article 7 of Directive 2012/27/ EU until 2020, the programme of renovation (modernization) of the multi-apartment buildings was amended by Resolution No 213 of the Government of the Republic of Lithuania of 25 February 2015. The main objectives of the mentioned programme in 2014–2017 are:

1. to ensure the financing and implementation of projects of renovation (modernization) of multi-apartment buildings meeting the requirements of the Multi-Apartment Building Renovation (Modernization) Programme – to grant preferential credits and other state support to owners of apartments and other premises, to promote the initiative of owners of apartments and other premises to implement energy-saving measures;

2. to develop public information, education and training on the issues of improving the energy performance of buildings, their renovation (modernization) and energy savings.

In 2017 alone, 403 multi-apartment buildings were renovated. According to the data on the monitoring of the implementation of the Multi-Apartment Renovation (Modernization) Programme submitted by the Ministry of Environment of the Republic of Lithuania, 403 apartment buildings were renovated. According to the data provided, the energy savings achieved in 2017 were 119.97 GWh. Building renovation facilities have a lifetime of 20 years. The total energy saving from 2014 to 2017 (until 2020) was 2525.33 GWh. According to the data provided by the Ministry of Environment of the Republic of Lithuania, in 2017, 762 measures were taken in implementing the second objective of the Programme and the amount of energy saved was 7.54 GWh. The period of life of information, education and training measures to improve the energy performance of buildings, renovation (modernization) and energy-saving measures is one year. The total energy saving from 2014 to 2017 was 62.39 GWh (ESTEP, 2019a).

On 3 September 2015, the Minister of Energy of the Republic of Lithuania and Minister of Finance passed order "On the Approval of the Rules of Implementation of the Ignalina Programme in Lithuania for 2014–2020". In accordance with these rules, a programme for the implementation of the EU financial support measures related to the decommissioning of Units 1 and 2 of the Ignalina Nuclear Power Plant in Lithuania (hereinafter referred to as the Ignalina Programme) was financed. One of the key focus areas of the Ignalina Programme was the renovation of multi-apartment buildings in the Ignalina Nuclear Power Plant region. According to the information provided by Ignalina, Visaginas and Zarasai district municipalities on the renovation of multi-apartment buildings in 2014, energy savings were 2.00 GWh in 2014, 1.62 GWh in 2015 and 1.63 GWh in 2016. Building renovation facilities have a lifetime of 20 years. The total energy saving under the Ignalina Programme between 2014 and 2017 (until 2020) was 36.43 GWh (ESTEP, 2019a).

The total energy savings under the Multi-Apartment Building Renovation (Modernization) and Ignalina Programmes are shown in Table 3.4. The table provides information on the results of energy efficiency improvement policy measures. The highlighted figures refer to the result of the measure – energy savings – in a given year. Depending on the duration of the measure in years, the result of the measure is felt either for only one year (information, consultation measures designed to change energy users' habits or the impact of excise duties and taxes on fuels for reducing fuel consumption) or many years ahead (physical measures for energy users). If the lifespan of a measure is more than one year, its energy savings shall be multiplied by the number of years up to 2020.

TABLE 3.4

Energy-saving targets, GWh

Measure	Energy savings, GWh							
	2014	2015	2016	2017	2018	2019	2020	Total
Multi-apartment building renovation (modernization) programme, Objective 1	25.30	25.30	25.30	25.30	25.30	25.30	25.30	177.10
		138.00	138.00	138.00	138.00	138.00	138.00	828.00
			208.07	208.07	208.07	208.07	208.07	1,040.35
				119.97	119.97	119.97	119.97	479.88
Multi-apartment building renovation (modernization) programme, Objective 2	41.12	6.45	7.28	7.54	—	—	—	62.39
Ignalina Programme	2.00	2.00	2.00	2.00	2.00	2.00	2.00	14.00
		1.62	1.62	1.62	1.62	1.62	1.62	9.72
			1.63	1.63	1.63	1.63	1.63	8.15
				1.14	1.14	1.14	1.14	4.56
							Total:	2,624.15

Source: Created by authors based on the Ministry of Environment of the Republic of Lithuania (2020).

The Multi-Apartment Renovation Programme has two main objectives:

1. to ensure the financing and implementation of renovation projects of multi-apartment buildings (modernization), i.e. granting preferential loans and other state aid provided for by law to the owners of apartments and other premises and encouraging them to take initiatives to implement the energy-saving measures;
2. to develop public information, education and training on improving the energy performance of buildings, their renovation (modernization) and energy savings.

To achieve these objectives, the programme set out the following quantifiable targets by 2020 (ESTEP, 2019a):

Objective 1: To renovate at least 4,000 multi-apartment buildings;

Objective 2: To raise awareness of the opportunities offered by the programme to the population of multi-apartment buildings by 90%;

- The number of people who want to participate in the implementation of the programme will be increased by 60%;
- Reducing the total energy consumption to 1,000 GWh per year by 2020.

Within the framework of this programme, all buildings built before 1993 can be financed.

The following are the measures qualified for financing renovation: insulating walls and roofs; replacement of windows and entrance doors; modernization of heating systems; renovation of ventilation systems or installation of recuperation systems; installation of alternative energy systems (sun, geothermal); glazing of balconies; and other systems such as elevators, electrical systems and hot/cold water supply systems. The average duration of the renovation process itself is 24 months. Before joining the programme, the energy certificate must be obtained, which will contain data on the energy consumption of multi-apartment buildings in the last three years. In Lithuania, 99% of multi-apartment buildings require complex energy efficiency measures such as wall insulation, roof insulation, window replacement and modernization of heating systems. Although monitoring of the implementation of this programme was carried out by the Ministry of Environment, the municipalities were expected to play an important role in achieving the main objectives of the programme. It is expected that municipalities will prepare, approve and supervise the implementation of municipal programmes according to the goals and objectives of the programme, coordinate the process of renovation of multi-apartment buildings in the municipal territory, linking it to the planning, renewal and management of the territory of the municipality (ESTEP, 2019a).

As regards the funding mechanism, the programme has been amended several times. Initially, the programme offered commercial loans, providing up to 50% of public grants. However, as a result of government budgetary constraints following the financial crisis, the level of government subsidies available has been reduced to 15%. Moreover, commercial banks were not very willing to risk loans for renovation. Since 2010, a financial instrument from the JESSICA I programme has been introduced and the owners of apartments have been offered a fixed interest rate of 3% on loans. In addition, a provisional repayment period of 10–20 years was established. The JESSICA II Fund was established in May 2015, which took over JESSICA I funding and used funds from the European structural and investment funds. One important priority of JESSICA II was to maximize private funding in order to minimize public funding. JESSICA II, unlike its predecessor, attracted more funds from commercial banks to support the programme. In 2015 "Šiaulių bankas" contributed to the programme €30 million (Skema & Dzenajaviciene, 2017).

In practice, apartment owners are offered a grant of 100% to prepare technical and financial documentation for renovation projects. However, the level of loan rebates varies according to the energy efficiency label achieved by renovation projects. By 2015, a discount of 45% was granted in order to achieve savings of at least 20% and to achieve energy efficiency class D. The rebate rate of the loan subsequently changed in 2015 and 2018, taking into account the financial conditions of JESSICA II. From 2015 to 2018 a discount of 40% of the loan was granted in order to achieve energy efficiency savings of at least 15% and to achieve energy efficiency class C. However, from 2018 onwards, a 35% discount on the loan was proposed in order to save at least 15% of energy efficiency class B.

All financial plans for renovation projects are designed in such a way that the repayment of the loan should not exceed 40–50% of the heating costs before renovation. By the end of 2019, 3,158 renovation projects in multi-apartment buildings were supported under this programme (Table 3.5). Further 1,609 projects are currently under way and are expected to be completed by 2021. The original goal of the programme was at least 4,000 renovation projects implemented by 2020. The figure reached so far set by the deadline is slightly above 75% of the initial target. However, when the number of completed and ongoing projects is aggregated (although 1,609 will be delivered a year later than expected), the final number of programme-supported renovation projects is 4,767. This means that the programme will achieve a surplus of 19% through renovation projects (European Commission, 2019).

As far as the total energy consumption is concerned, the programme has so far achieved energy savings of 700.76 GWh, with an overall estimated savings of 2.3 TWh, which is less than the primary target of 1,000 GWh per year by 2020.

There are several reasons why the original goals will not be achieved in time. The main reason is that the 2004 targets were too ambitious given the

TABLE 3.5

Renovated multi-apartment buildings, 2005–2019

Year	Renovated multi-apartment buildings
2005–2011	822
2012	31
2013	41
2014	123
2015	574
2016	769
2017	403
2018	224
2019	171
Total	3,158

Source: Created by authors based on European Commission (2019).

limited capacity of the state to subsidise renovation projects. Although the first renovation projects under the programme were launched in 2005, due to insufficient funding, renovation activities were very limited. Between 2005 and 2010, only 375 apartment buildings were renovated. Most renovation activities in Lithuania started to grow only in 2010, with the introduction of the JESSICA financing mechanism. For example, in Kaunas municipality, the second largest municipality in Lithuania, the first projects started only when the municipality received financial support from EU funds (Skema & Dzenajaviciene, 2017).

Renovation projects implemented by 2020 would exceed 75% of the original targets. Many projects are still ongoing and are expected to be completed in 2021. After the implementation of the programme, the renovation project will have to reach a surplus of 19%, albeit one year later than planned. It can, therefore, be said that the measure has been successful; however, there is still much room for improvement. Given the total number of multi-apartment buildings that still need to be renovated, there are a number of areas for the implementation of the current programme which need improvement. For example, to assess the quality of renovation better maintenance is needed, better and more reliable data on energy use/consumption are needed to carry out energy audits and more diversified financial support measures are needed to encourage more homeowners to undertake renovation work. Yet, the programme should exceed its initial goals by 19% supporting 767 renovation projects additionally than planned. The implementation of the new governance model and the JESSICA II funding mechanism has improved significantly. As a result, municipalities have become more actively involved in the programme and now more apartment owners are involved in the programme (European Commission, 2019).

The following recommendations are possible (European Commission, 2019):

- to continue close cooperation with municipalities by improving the project identification, selection implementation and evaluation process and coordinating with the administrators of the renovation project; to explore ways of using municipal borrowing opportunities and seeking simplified contractual arrangements;
- to provide support for multi-apartment owners in taking collective decisions to help them agree on investments and implementing decisions; a simple majority should be sufficient to make the decisions by homeowner's association on borrowing and contracting;
- to establish a strong central centre of excellence to advise house owners and assist them in the renovation process; this support measure would not only facilitate the process of beneficiaries but would attract more involvement;
- to ensure that renovation projects are carried out in accordance with good quality standards, effective use of technical standards, maintenance of works and quality assessments, and that sufficient and reliable data are available to carry out energy audits and to assess the need for and expected effects of renovation works; audits should assess the quality of building structures, heating and ventilation systems as well as microclimate parameters;
- to develop alternative public and private sector financial support programmes for the renovation of multi-apartment buildings, which could include the refurbishment of all micro-districts and infrastructure, together with the renovation of multi-apartment buildings, as a principle of future energy efficiency improvement programmes;

There are a number of important problems that have hampered the success of the programme. This measure would probably be best suited for countries or regions whose housing sectors are in a similar situation to Lithuania, e.g., the old building stock built during the Soviet period, which has not been properly maintained and is highly energy inefficient. The current application of the programme is highly dependent on EU funding and is likely to be most easily applied to those countries that have recently joined the EU or are in the process of accession. However, the sustainability of this type of instrument depends on the successful support of different sources of funding (public and private).

3.3.3 Achieved Energy Efficiency Improvements and the State of the Renovation of Multi-Apartment Buildings

In Lithuania, energy consumption per unit of gross domestic product in 2019 was 1.88 times higher, according to Eurostat data, than the EU average, showing low energy efficiency in the country. Energy efficiency in the country is measured on the basis of primary and final energy intensity, indicating how much energy costs were incurred to produce a country's GDP. In 2010–2019, the primary energy intensity in Lithuania decreased by almost 30% (ESTEP, 2019a).

Lithuania does not achieve its goal of increasing energy efficiency as the country's implementation of energy-saving measures is delayed. It was necessary to ensure that legal acts were adopted in Lithuania by 5 June 2014, enabling energy efficiency improvement in Lithuania and energy savings of at least 2.6–3 TWh in multi-apartment buildings and public buildings by 2020 and 5–6 TWh by 2030. Accordingly, alternative energy-saving measures were planned to save 5.64 TWh, or 48% of the target energy savings, i.e. energy savings of 2.44 TWh are foreseen in the agreements with energy suppliers on consumer education and advice, while energy-saving agreements with energy companies are expected to achieve savings of 3 TWh and savings of 0.2 TWh for boiler houses. The NEIS stipulates that more than 70% of multi-apartment buildings and most public buildings use heat in an inefficient way, and their slow modernization may lead to serious economic and social consequences. About 70% of Lithuanian residents live in multi-apartment buildings, almost all of which were built before 1993. Such multi-apartment buildings are energy inefficient and do not meet modern requirements and standards, and their engineering systems are completely worn out. Since 2013, only 2,088 multi-apartment buildings have been renovated in Lithuania. Even 90% of apartment buildings were built until 1991. There are 35,500 multi-apartment buildings in Lithuania. About 500 multi-apartment buildings are renovated in Lithuania every year. If the pace remains the same, all multi-apartment buildings will be renovated only after 70 years (ESTEP, 2019a).

Implementing the multi-apartment building renovation programme, over 300 multi-apartment buildings or 9,000 apartments were renovated in Lithuania last year. Since 2013, when municipalities were involved in the renovation process, more than 24,000 multi-apartment buildings or 700,000 apartments were renovated by 2019.

In order to improve the energy efficiency of the housing unit, it is necessary to prepare detailed neighbourhood plans combined with mobility plans, development and renovation plans of public buildings and infrastructure, i.e. in designing cities to project buildings whose useful area could be increased (increasing the number of floors or constructing extensions) where new buildings could emerge, etc. Renovating according to detailed neighbourhood plans can achieve economies of scale by reducing the cost of renovation and having positive external effects on other areas. In Lithuania, in most cases, multi-apartment buildings were only partially renovated, as residents independently implement the most important energy-saving and effective measures such as changing windows and doors. Meanwhile, a complex renovation of the house increases the payback period. The average payback

TABLE 3.6

Calculation of the repair period for the renovation of a multi-apartment building

Presumption	Value
Total area of a typical multi-apartment building, m²	1,287
Investments in euros/m²	195
Investments for the project, in euros	250,965
Current consumption in kWh/m²/year	195
Current annual building consumption, kWh	225,225
Heat energy price in €/kWh	0.053
Annual heat price growth	2%

Source: Created by authors based on ESTEP (2019a).

period of one renovation project is over 15 years. According to the survey of commercial banks, commercial banks grant investment loans for a period not exceeding ten years. As the duration of the loan is shorter than the payback time of the project, commercial banks do not provide such a loan, or the borrower must find other sources for servicing the loan, as energy savings are not sufficient. The assumptions for calculating the payback period for the renovation of multi-apartment buildings are shown in Table 3.6 (ESTEP, 2019a).

Currently, in Lithuania, energy consumption in residential housing is not taxed or taxed symbolically, except for the excise duty of €1.08 per MWh, which does not have a significant impact on the decisions of the population, especially when the use of gas in heat production has decreased significantly in the last ten years. Table 3.7 provides a sensitivity analysis for the multi-apartment renovation payback period in terms of energy savings and the amount of subsidy.

The research carried out in Lithuania has shown that the irrational behaviour of consumers in the field of energy efficiency is widespread in Lithuania, and a general perception of the importance of energy saving and energy efficiency has not yet been fully developed. Similarly, due to their low financial education, energy consumers often misjudge their energy consumption and savings and the period of return on investment and remain more likely to avoid investment and continue to pay more for their energy consumption than to take appropriate actions. Since 2013, when measures were initiated for the renovation of multi-apartment buildings, and even though residents used public investment more actively, it has not been understood whether the Lithuanian population's attitude towards energy efficiency investments has changed or whether residents have rushed to make use of subsidies introduced by the Ministry of Environment for the results of energy savings achieved, which collectively achieved up to 40%.

In Lithuania, energy consumers often do not have enough information and are not aware of the existing technologies for energy savings. Insufficient

TABLE 3.7

Sensitivity analysis of the payback period of multi-apartment renovation in terms of energy savings and amount of subsidy

Subsidy	Energy savings								
	40%		45%	50%	55%	60%	65%	70%	75%	80%
25%		68.8	50.1	40.2	33.8	29.3	25.9	23.2	21.1	19.3
30%		56.4	43.2	35.4	30.2	26.3	23.4	21.1	19.2	17.6
35%		47.3	37.5	31.2	26.8	23.6	21.191,11	19.0	17.4	16.0
40%		40.2	32.6	27.5	23.8	**21.1**	18.9	17.1	15.7	14.4
45%		34.3	28.3	24.1	**21.1**	18.7	16.8	15.3	14.1	13.0
50%		29.3	24.5	**21.1**	18.5	16.5	14.9	13.6	12.5	11.6
55%		24.9	**21.1**	18.3	16.1	14.4	13.1	12.0	11.0	10.2
60%		**21.1**	18.0	15.7	13.9	12.5	11.4	10.4	9.6	8.9
65%		17.6	15.1	13.3	**11.8**	10.7	9.7	8.9	8.2	7.7
70%		**14.4**	12.5	11.0	9.9	8.9	8.1	7.5	6.9	6.5

Source: Created by authors based on ESTEP (2019a).

investment in energy efficiency measures is also accompanied by a lack of information resulting from the lack of energy literacy and motivation to acquire it. Residents, who are final beneficiaries of multi-apartment modernization projects, are often unaware of the benefits of implementing energy efficiency measures, whereas the overall process of consensus between residents of multi-apartment buildings can be very long, and sometimes this consensus is not reached at all. For many other reasons, residents of multi-apartment buildings in Lithuania have low motivation to initiate energy efficiency projects themselves.

According to the Energy Efficiency Assessment Report, Lithuania needs to allocate more than €1.3 billion for the renovation of multi-apartment buildings over the 2014–2020 programming period in order to achieve the desired results, but taking into account all sources of financing available in the market, including ESF funds and co-financing for the renovation of multi-apartment buildings, is insufficient. In order to continue the activities of the JESSICA holding fund (JESSICA HF) created in 2009, around €280.9 million were allocated from the ESF for the renovation of multi-apartment buildings during the 2014–2020 period; the amount was then increased to €314 million. However, even with the additional assessment of the returned funds of JESSICA HF, the need for investment is more than a few times higher.

Between 2014 and 2017, €513 million was invested under the Multi-Apartment Building Renovation Programme. Given the remaining €87.8 million in the multi-apartment renovation funds in 2018, it is likely that these funds will not be sufficient to renovate 500 multi-apartment buildings per year for the fulfilment of the objectives of the programme, as €150 million per year should be allocated based on the number of applications received.

TABLE 3.8

Demand, supply and delay for building renovation investments

Type of buildings	Demand for investment, € million	Investment supply, € million	Lack of investment, € million
Central modernization of government buildings	610.3	109.8	500.5
Modernization of municipal buildings	298.6	17.3	281.3
Renovation of multi-apartment buildings	1,050	513	449

Source: Created by authors based on ESTEP (2019a).

Renovation programmes for multi-apartment buildings and public buildings are expected to generate energy savings of 2.67 TWh by the end of 2020 and savings of 2.1 TWh by the end of 2016. One of the most serious reasons for inefficient energy consumption in Lithuania is the very poor thermal properties of most public buildings requiring a lot of energy to heat them. Environmentally sound RES such as geothermal, solar and wind energy are under-used to generate heat and electricity. The housing and transport sectors take the largest reserves for improving energy efficiency. Residential house renovation would save a lot of energy and significantly reduce GHG emissions, but despite many years of efforts, renovation in Lithuania did not become mass.

In Lithuania, modernization, renovation and energy savings of buildings require significant investments (Table 3.8).

According to the data of the Lithuanian Department of Statistics, thermal energy is used mainly for heating housing. Heat consumption increased by 5% between 2013 and 2017, from 4,459 GWh to 4,661 GWh. The use of natural gas in oil heating also increased by 32% between 2013 and 2017 and electricity by 12%, while the use of wood for fuel and agricultural waste decreased by 12% during the same period (Table 3.9).

It has been found that energy efficiency in both the residential and public buildings and transport sectors is determined by detailed neighbourhood plans combined with mobility plans, i.e. the most energy efficiency can be achieved by planning urban development and anticipating appropriate changes in buildings and infrastructure. For example, planning the increase in population density in the quarter, providing schools, kindergartens, shops and other necessary infrastructure for the population would reduce the need to travel.

At the same time, the replacement of heating systems in all these buildings into low-temperature heating systems, the renovation of heating transmission routes, can lead to higher energy efficiency with minimal investment. Public investment will also be better used in anticipating the development of the electro mobile loading infrastructure. This would help to avoid

TABLE 3.9

Type of energy used for house heating

Fuel used for house heating housing	2013	2014	2015	2016	2017
Wood, thousand m³	5,785.8	5,468.0	5,295.2	5,261.7	5,101.3
Natural gas, GWh	988.2	970.1	995.6	1,216.1	1,299.5
Electricity, GWh	134.8	140.8	141.0	141.5	150.4
Other fuels, GWh	751.5	661.1	542.7	632.1	740.6
Thermal energy, GWh	4,458.5	4,175.7	4,173.2	4,599.2	4,660.5
Heat demand, total GWh	12,118.7	11,415.8	11,147.6	11,850.5	11,952.4

Source: Created by authors based on ESTEP (2019a).

over-investment when new buildings or infrastructure are renovated or constructed; however, later, they are rendered redundant or insufficient.

Well-developed district heating systems operate in all cities of Lithuania, providing heat to about 53% throughout the all country and in cities – about 76% of all buildings. A total of 18,177 (about 700,000 apartments) out of 27,775 district heating buildings are multi-apartment buildings. Accordingly, district heating is an important element in achieving energy efficiency; nevertheless, in addition to district renovation and conversion of heating systems into low-temperature districts, it is difficult to achieve greater energy efficiency in major cities.

In 2017, biofuels accounted for 65% of district heating production. It is planned that in 2020, biofuel heating in the district heating system will grow to 80%. It is also important to note that the introduction of biofuels that are more expensive than natural gas has led to a 20–45% reduction in heating costs in urban areas. Such cities as Kaunas, Pakruojis, Utena, Plungė, Taurage, Šakiai, Varėna, Molėtai, Ignalina, Lazdijai, Skuodas and others produce more than 90% of heat from biofuels. The largest city, Vilnius, produced on average 45% of heat from biofuels that year, and by 2020 it is planned to reach 80%.

However, the increased use of biofuels and falling heat prices have also reduced the incentives and benefits of the introduction of energy efficiency measures and the renovation of buildings, respectively; in order to speed up the pace of renovation, additional measures such as increasing heat prices by terminating tax incentives, imposing additional taxes or reducing the cost of renovation through tax incentives and stimulating growth in the renovation market and competition in the sector are necessary.

3.4 Conclusions

Lithuania has set ambitious RES development goals in 2020, 2030 and 2050 final energy consumption (30%, 45% and 80%, respectively) and electricity

(0%, 45% and 100%, respectively), transport (10%, 15% and 50%, respectively) and heating and cooling sectors. The RES development in Lithuania is carried out taking into account EU and national strategic documents and legal acts. The main RES development policies and measures are embedded in the updated NEIS, the National Strategy for Climate Change Management Policy, the Law on Energy from Renewable Sources of the Republic of Lithuania and integrated into the National Energy and Climate Action Plan of the Republic of Lithuania 2021–2030.

It should be noted that the full implementation of the planned policy measures in all sectors is required in order to achieve the targets set for the promotion of RES in 2030. One of the most important directions in the field of RES is the promotion of active electricity consumers who are able to consume electricity for their own needs. Such active consumers must receive a remuneration appropriate to market conditions for the surplus energy supplied to the grid. Lithuania also aims to promote the active participation of local communities in investing in RES facilities managed by joint ownership.

The goals of energy efficiency and renovation of buildings are detailed in the National Action Plan for 2014–2020, the National Environmental Strategy, the National Strategy for Climate Change Management Policy, the Energy Efficiency Improvement Action Plan for 2017–2019, the Interinstitutional Action Plan for implementation of goals and objectives of the National Strategy for Climate Change Management Policy for 2013–2020. These documents set goals for energy efficiency and renovation of housings in various ways.

Lithuania plans to save 11 TWh of energy by 2020. Energy saving for renovation of multi-apartment buildings was 2.7 TWh, while energy saving for renovation of public buildings was 0.4 TWh. Energy savings of 2.4 TWh are planned to be achieved by informing consumers. The main problems related to the renovation of multi-apartment buildings are the payback period for renovations, the absence of single neighbourhood renovation plans, the volatile flow of projects and insufficient funds for renovation. In addition, tax-free household pollution and asymmetric information and consumer behavioural patterns of energy savings are key barriers to improving energy efficiency in this sector.

The programme for renovation (modernization) of multi-apartment buildings, approved in 2015, set up energy savings to 2020 goals and aimed at ensuring preferential credits and other state support for owners of apartments and other premises, promoting the initiatives of owners of apartments and other premises to implement energy-saving measures and to develop public information, education and training on the issues of energy efficiency improvement, renovation (modernization) of buildings and energy savings.

The results of the programme show that renovation projects implemented by 2020 will reach 75% of the original goals. Many projects are still ongoing and are expected to be completed in 2021. After the implementation of the programme, the renovation project will have to reach a surplus of 19%, albeit one year later than planned. It can, therefore, be said that the measure has been successful.

Given the total number of multi-apartment buildings that still need to be renovated, there are a number of areas for the implementation of the current programme that need improvement. For example, better maintenance is needed to assess the quality of renovation, better and more reliable data on energy use/consumption are needed to carry out energy audits and more diversified financial support measures are needed to encourage more house owners to undertake renovation works.

Seeing that primary and final energy consumption have been increasing slightly over the last few years and seeking to properly implement the set targets for 2020, Lithuania is undertaking additional energy efficiency improvement measures. During 2019, a financial measure was introduced to encourage residents to replace inefficient heating installations (biofuel boilers) with more efficient technologies that use energy from renewable sources (implemented in 2019).

Lithuania has developed the National Energy and Climate Action Plan for 2021–2030 in line with the requirements of the Energy Union Governance Regulation. The National Plan has been prepared on the basis of and integrating the provisions of Lithuania's national legal acts, international obligations, strategies and other planning documents, objectives and tasks as well as the implemented and measures planned to be implemented. The main strategic documents integrated into the National Plan are the NEIS approved in June 2018 and the National Strategy for Climate Change Management Policy approved in 2012 and updated in 2019 as well as the National Plan for the Reduction of Air Pollution approved in April 2019.

4

Barriers to Climate Change Mitigation in the Energy Sector

4.1 Market Failures Preventing the Introduction of Climate Change Mitigation Measures

There are the following generally recognized market failures: external effects, public goods, incomplete markets, imperfect competition, imperfect information, uneven distribution of income, unemployment and other macro-economic shocks. Some market failures overlap (Bator, 1958). Information problems often explain part of the non-existing market failures. External effects are often caused by non-existing markets, whereas public goods are sometimes considered to be extreme external effects, where others derive the same benefits from the manufactured product as the manufacturer itself (Stiglitz, 1989).

External effects. Environmental economic theory has an important role to play in the concept of external influences or external effects. External effects are a special case in the incomplete market of natural capital. If the consumption or production of a single individual or firm affects the benefit of another person or the function of the firm's production, when Pareto-optimal conditions of allocation of resources are violated, there are external effects. We will find that this external effect does not manifest itself through market prices, but through its effect on benefits or profits. The set of markets is incomplete, as there is no exchange authority where a person pays for external benefits or for causing external effects. External effects (costs or benefits) are effects on third party income or costs that are external to the market, i.e. they are not reflected in market prices. They are experienced by individuals who do not participate in market deals. In the energy sector, such negative external effects result from atmospheric pollution, energy production or consumption of energy products, including the cost of climate change, positive externalities due to the introduction of renewable energy sources (RES) and energy efficiency improvement measures and external costs resulting from a reduction in the reliability of energy supply. Energy production and consumption also lead to other negative external effects related to chemical and thermal pollution of water, external costs of oil products entering into water and land, external costs of using RES such as land inundations, landscape deformation, noise caused by the wind power plant and particulate matter emissions due to biomass combustion (Medema, 2007).

All these external costs reflect the impact by which the optimal level of reduction can be determined. External effects are a specific case of a lack of markets. Pollution leads to costs for those who suffer harm as a result of pollution, and they are willing to pay a certain amount to avoid this pollution; however, there is a lack of a market where they can express their willingness to pay. Generally, the readiness to pay is less than the avoided small amounts and is growing as the latter grow. Reducing pollution is usually costlier. It is cheaper to reduce significant increases in pollution and to reduce the increase in lower pollution at a higher cost. Depending on the nature of the external effect, the situation can be illustrated by a decreasing environmental quality curve and an increasing environmental quality supply curve. Where there is a market, polluters and population affected by pollution may agree that the costs of reducing the last unit of pollution increase should be equal to the preparedness of the affected population to pay for the last liquidation of pollution increase. When the optimal level of pollution reduction is established, polluters can easily be taxed at a level equal to the marginal cost of the reduction compared to the marginal damage to the society (Krugman & Wells, 2006).

Since a clean environment is a public good, society can set external costs and impose a polluter tax on polluters. The polluter is thus forced to include external costs in the cost of production. As a result, pollution decreases, the cost of producing goods increases and the consumption of goods in response to higher prices is reduced. However, it is possible that people affected by pollution, knowing that they will not have to pay for reducing pollution, may overestimate their hypothetical willingness to pay.

In competitive markets, producers seek to minimize production costs in order to reduce the cost of their production, increase production volumes and improve the quality of their products. Consumers demonstrate how much they are willing to pay for the product and how much they want to buy. Given the equilibrium price in competitive markets, specific efficiency is ensured. When the marginal benefits of a product exceed the marginal costs, the situation can be improved by increasing the quantity of this product. The influence of exchanges between buyers and sellers on other persons who are not directly involved in the trade produces external effects. External effects are negative or negative effects of described exchanges for persons not involved in market transactions. These are the effects on a third party. When these effects are pleasant or beneficial, they are called positive external effects or external benefits. When these effects are unpleasant or cause costs to people who are neither buyers nor sellers, they are called negative external effects or external costs (Krugman & Wells, 2006).

Ideal competition in markets where all players operate independently of each other may force producers to pass on external costs to others. The production costs of a person who voluntarily install purification equipment will be higher than those of its competitors, so the person will be forced to leave a business in the long term without being able to reduce prices. As a result, many manufacturers ignore the fact that they cause harm to others by polluting the environment. Even socially responsible producers are sometimes forced to neglect the environment as much as they should.

Other examples of external costs, which leads to overproduction, include the congestion on highways, aircraft noise caused in the airport territory, the deaths caused by speeding and irresponsible driving. If the State did not regulate driving and noise norms, there would be high external costs for third parties. An example of external costs is the costs of sex shops and massage salons for neighbourhood businesses, as the neighbourhood of such institutions discourage those who want to visit their shops selling normal goods or companies providing conventional services (Bator, 1958).

Some analysts first noted that pollution cost problems are not widespread and can be adequately solved by complementing the existing ownership right structure. (Modern English and American ownership rights sanction individual ownership of almost all things that have economic value). It can be shown that when ownership rights are precisely defined, individuals and firms have incentives to use natural resources as efficiently as possible. Moreover, supporters of the ownership rights approach argue that with adequate identification of ownership rights, the public good – the quality of the environment – can be transformed into a private good and thus the optimal distribution of the environment can be achieved. State intervention is sometimes necessary to determine who owns environmental property rights. And then the market itself will find a fair solution for the allocation of resources (Medema, 2007).

Environmental pollution is one of the forms of market failure often caused by the re-exploitation of resources when they are jointly owned or without an owner at all. Thus, the market is "trapped" when ownership rights are not clearly defined or uncontrolled by those who could personally benefit from using resources in the most efficient way (Stiglitz, 1989).

Market demand does not always reflect the benefits derived from the good. People who do not participate in market transactions can sometimes receive part of the benefits of the product free of charge. The use of RES can be an example of such external benefits. Due to the use of these resources in the production of energy the negative impact on the environment is reduced, fossil fuel resources are saved, new technologies are adapted and so on. All of this benefits those who do not participate in green energy production or those who do not consume. Since the free market is unable to assess these external benefits, the State must subsidize green energy. Another example is vaccination, where the external benefits outweigh the private benefits for the one who is vaccinated; hence, the State usually subsidizes not only vaccines but also research programmes in this area. All public transport, public parks and the informed and literate population receive external benefits. All these goods are characterized by high costs and therefore cannot be fully realized by private funds. Additional funds from the State budget are needed to replenish private funds and encourage private services (Medema, 2007).

Marginal social costs of production are below marginal social benefits. In this case, social marginal costs may be lower than marginal social benefits. For example, due to the job-creation effect, the external effects of the use of RES are due to the positive environmental impact of fuel oil replacement by RES. Let us consider the use of petrol, which has a negative impact on society as a result of environmental pollution. The last litre of petrol undoubtedly gives an advantage equal to its price (litre gasoline) to the individual consumer of petrol, who pays a certain price. Otherwise, an individual would not buy it. However, for the surrounding people in the city the last consumed petrol litre value is negative due to carbon emissions (Stiglitz, 1989).

Incomplete markets. When private markets fail to deliver goods or services, although the cost of producing them is lower than the price people are willing to pay, there is a market failure that we call an incomplete market (as a complete market would deliver all goods and services whose production costs are lower than the price people are willing to pay). The private market does not cover many important risks (e.g. deposit insurance, farmers' harvest insurance, mortgage loans, student and small business loans, health insurance for the elderly, unemployment insurance, insurance against terrorism, natural cataclysms). The question – why the capital and insurance markets are imperfect – has been very much discussed over the last two decades. There were at least three different answers, and each one is correct in its own right. One focused on innovation: we are accustomed to the emergence of new products. However, there is also innovation and economic development: innovation in creating new markets, including the emergence of new securities and new types of insurance. These new products are related to the second explanation – transaction costs. It is expensive to create a market and a new product. An insurance undertaking may be reluctant to do so if it is not certain that there will be a sufficient number of customers. There is no effective "patent protection" and the result will be an insufficient investment in innovation. The third explanation is based on information asymmetry and enforcement costs. The insurance company is often less informed about the nature of the client's risks than the client himself. When the two parties of the transaction have different information, we say there is information asymmetry (Krugman & Wells, 2006).

In this way, the peasant may want to buy insurance for bad harvests, but the insurance undertaking itself will assess the risk and, accordingly, apply for a certain amount of insurance premium. But if it overestimates the risk, the bonus will be too high, and the peasant will not buy a policy. If the insurance undertaking underestimates the risk, the premium will be too low: a peasant will buy a policy, but on average, the insurance undertaking will lose money. When similar information asymmetry is high, markets do not exist.

Similarly, in the capital market, lenders are concerned to get their money back. They can hardly distinguish which borrowers are likely to repay the loan. The bank faces a dilemma: if it raises the interest rate to cover losses due to outstanding loans, it may notice that insolvency actually increases because those who know that they are going to repay are refusing to borrow, and those who do not plan to repay are borrowing at all costs because they will not repay. This phenomenon is called an adverse selection. It plays an important role in the health insurance market. The problem of asymmetric information and enforcement costs explains why there are few many markets. The reasons why markets do not exist can have consequences for how governments treat market failures. The government also has transaction costs, the problem of enforcement costs and information asymmetry, although in many cases they differ from the respective private sector problems. In this way, when developing loan programmes or interfering in capital markets, governments must take into consideration that they are also often less informed than borrowers. Therefore, there is a problem with the absence of certain complementary markets. Suppose there is no sugar market without the coffee market. If two businessmen agreed, there would be a market for both coffee and sugar. Everyone acting alone could not pursue the public interest. There are many cases where there is a need for very complex coordination among many people, especially in developing countries, and may require state planning. If the markets were full, prices would perform this "coordination" function (Krugman & Wells, 2006).

Public goods. What nature supplies us – the air to breathe, the amenities of the landscape, the recreational function of nature and various ecosystems – are all public or public natural consumption goods. We know from the course of economic theory that public goods can be described in two ways (Bator, 1958):

1. Unlike private goods, the public good can be used at the same time by several individuals who do not compete with each other. Classifying a product as a public good is not a sufficient condition for collective consumption, collective consumption can also be characteristic of many private goods such as professional boxing competitions and other mass entertainment events.
2. However, the public good prevents the exclusion of competing consumers (as opposed to watching sports competitions or other mass entertainment events). The most suitable example here is the classic lighthouse, the light of which is used by all ships at sea, regardless of whether or not they contributed to the construction of the lighthouse.

A distinction must be made between private goods, the supply of which will be guaranteed by the private sector because they will be able to generate profits, and public goods, which will not be guaranteed by the private sector since the latter will not be able to profit from it. Public or common goods are characterized by two features: the consumption of a public good does not exclude other consumers from consumption; potential consumers will, therefore, try to avoid paying for that good and consumption of the good by other additional consumers does not reduce the consumption of the good for other consumers, i.e. the marginal social cost of supply is zero. On the contrary, private goods have well-defined ownership rights (thus separation of other consumers is possible) and their marginal supply costs are increasing. Thus, markets can ensure an optimal supply of private goods, but they cannot ensure an optimal supply of public goods. In addition, in the case of public goods, there are serious problems in assessing the readiness to pay, as consumers are characterized by reduced readiness to pay, i.e. evading payment from those who declare their willingness to pay and then do not pay for it. Public goods include valuable products such as national defence and social infrastructure. This includes fundamental research necessary to ensure the development of energy efficiency improvement measures and RES. The private sector will, therefore, be reluctant to fund fundamental research and public support is needed to ensure such research (Krugman & Wells, 2006).

Let us look at the condition of separating public goods from the impossibility of consumption. For example, the development of a new energy-using product requires fundamental research on materials. A company carrying out fundamental research will notice that if critical phenomena are discovered, other firms will use their achievements in their products. These forms will then be able to compete with the inventors' firm, without any inclusion of research costs in the cost of their product, since they did not do so. Thus, the inventors' firm will not be able to cover the costs of research by including the cost of their production and will, therefore, have no incentive to carry out such fundamental research at its own expense.

There are several possibilities to limit the use of ideas developed by the firm and to allow undertakings to carry out fundamental research. For example, patent registration gives intellectual property rights to the research firm and allows the costs of research to be covered, as other undertakings will have to buy a patent to make use of the results of the undertaking's research. However, it is important to set a threshold that distinguishes between ideas that should be protected through patents and ideas that should not be protected by patents. Under the zero marginal cost clause, no firm will use ideas to reduce the knowledge of other firms and it can be claimed that the marginal costs of knowledge and information are equal to zero. Thus, in addition to patent protection, all research and development must be at the disposal of the State and the non-profit sector. Moreover, it is common knowledge that firms have more experience than governments in meeting market needs in the most efficient way; therefore, it is desirable that firms carry out applied research and development activities to ensure the supply of private goods to the markets. This makes it possible to lay down guidelines in patent policy. Patents cannot be granted for ideas. They come from inventions that are unique, have direct benefits in terms of application and are not obvious. In other words, patents are issued for innovation with practical results and improve the current situation in a meaningful way. The inventor must repay for patent protection by providing a written description of the invention and instructions on how to use it. When an invention is used in the manner described in the patent, the inventor must receive compensation and, where the invention simply enriches knowledge in a particular area, no compensation is paid to the inventor. The exact application of the patent law to an individual product varies according to the product and depends on applicable laws, guidelines and the State policy. The basic rule in patent policy is as follows: the more fundamental the invention, the more State support it needs. The more the invention is linked to the creation of a product that is attractive to the market, the less public support is needed for such research and the greater the private sector's interest in such research is. It follows that the private sector assesses the development of a new product on the basis of its potential attractiveness on the market (Krugman & Wells, 2006).

The nature of training and education as a public good is an important basis for government involvement in energy efficiency and renewable energy development programmes. Investment by private firms in the education and up-skilling of employees is hampered by fears that an employee can change jobs and not work all the time that would allow the money spent on training to pay off. The difficulties and complex process faced by firms when installing energy efficiency improvement equipment, compared to simple energy purchase, can hamper many cost-effective investments. This is a particularly high barrier to small- and medium-sized undertakings. In some firms, there is a serious lack of trained technical staff to understand and explain the potential of energy-saving technologies to save money instead of the installation of new additional powers. State-funded programmes, where university staff are paid for energy audits in industrial undertakings, can help overcome this energy market shortage by training the next generation of energy professionals and providing energy diagnostic and audit recommendations to factory managers (Medema, 2007).

In the case of public goods, individual consumers cannot choose different quantities of this product that they consume independently of each other. If one individual consumer consumes an appropriate quantity of a public good, every other must use the same quantity precisely. As far as public goods are concerned, another important case is national security, which relates to the supply of oil products. After all, national security is a classic social good. No individual importer of petroleum products correctly reflects national security interests by deciding how much to import. Thus, leaving the market to decide how much it can import and how much oil it produces, there is an excessive dependency on import. Oil markets also impose untaxed national security costs. Under relatively strict market conditions, the physical concentration of oil reserves in relatively few countries poses a threat to physical and price-based energy supply interruptions. Such market conditions generate national security costs by reducing the flexibility of foreign policy and complicating militaristic strategies, particularly in periods of oil demand growth and global market constraint. In addition, the problem of greenhouse gas (GHG) emissions requires investment in GHG reduction technologies, although the cost of today will also benefit future generations. Thus, energy efficiency solutions arc based on the readiness to pay to raise the validity of justice decisions as well as even greater challenges such as equity in the distribution of costs and benefits between generations (Krugman & Wells, 2006).

In summary, the role of the State is expressed in addressing the efficiency challenge between the limitation of resources and the readiness to pay in order to ensure the highest economic well-being at a certain level of allocation of economic resources.

The criterion of justice or moral obligations regarding the supply of resources to those with limited initial resources is not assessed. The criterion of justice is of great importance in politics. Moreover, energy is valued not by itself but as the services it provides. Thus, fuel oil and other petroleum products are valued for the assessment of domestic heating, whereas petrol is assessed to evaluate the benefits of cars.

However, by regulating car fuel consumption standards and improving fuel efficiency, we are facing an increase in the use of cars, which, in turn, may lead to additional negative external effects. Monitoring changes in the car use patterns resulting from fuel efficiency policies can identify external costs due to the increase in car congestion or accidents, but not as a result of an increase in emissions into the atmosphere. To summarize, it can be said that private action has an impact on the consumption of shared resources, but there are many pertinent circumstances regarding the balancing of the costs and benefits of these actions. External cost assessments and the public's willingness to pay are difficult to assess and there is clear disagreement among scientists on which assessments could be considered the best. There is also no clear consensus on which incompatibilities or imbalances are important (Krugman & Wells, 2006).

There is a logical correlation between a greater fuel economy and a reduction in GHG emissions, together with an increase in car consumption and higher traffic jams and accidents. However, many find it difficult to assess the cost of controlling GHG emissions as the cost of increased traffic jams. Thus, in the assessment of external costs and in the fight against other market failures, the issue of choice remains a very important issue (Medema, 2007).

Monopolies and falling costs. The formation of monopolies is one of the most important examples of the limitations of the market mechanism. The market mechanism in itself creates a monopoly, which, in turn, reduces the efficiency of public production. In the case of a natural monopoly, the large undertaking gains market power because, in certain areas of activity, economies of scale result in a much lower cost per undertaking than a few competitors. Thus, where there is only one firm on the market, it has an influence on the market price and it can choose the price and the volume of production that will maximize its profits. However, in reality, the monopoly cannot choose between price and quantity of production completely free: it will sell only as much production as demand will be. If the monopoly imposes a high price, it will be able to sell only a small quantity of production. The demand resulting from consumer behaviour will determine the price and the volume of production chosen by the monopolist. It is possible that the monopoly chooses the price, after which consumers determine how much they would like to buy at this price; alternatively, a monopoly chooses the amount of production it produces, and then consumers decide how much they can pay for this quantity. The first condition is more natural, though the second condition is more appropriate for analysis. In fact, both conditions are equivalent when the analysis is correct. The monopoly is characterized by a concentration of supply in one hand. An absolute or pure monopoly is a type of market structure where it is the only seller of goods without close substitutes. This term also refers to the sole seller or producer of the goods.

Since the product produced or sold by the monopoly is unique in that it does not have close substitutes, there are no alternatives for the consumer: he has to buy from a monopoly or manage without it. If there are substitutes, then advertising is needed. For example, a consumer can buy a diamond or spend money for rest at the sea. Convincing advertising will help the consumer to make a decision. Domestic goods and services such as water are not substitutes; they do not need advertising. A pure monopoly is an abstraction. There are very few (if any) products in real life that do not have substitutes. The local electricity company in the city may be the only distributor of electricity in the region, but electricity, as it is used for various purposes, has alternatives. If the price of energy increases, the heating electricity is replaced by coal. When a single large firm, meeting all market demand, can produce a whole product with lower average costs than other competing firms, such a firm may, by lowering its prices, drive out of the branch small firms in order to meet all market demand. A firm that can meet all market demand with lower average costs than the costs if two or more firms would produce the same volume of production would be is called a natural monopoly (Medema, 2007).

Natural monopolies exist in areas such as indigenous gas supplies and electricity. The economies of scale of production are the main reasons why firms enjoy monopolistic rights. Changes in technology can change cost conditions. For example, long-distance telephone communication was once a natural monopoly, but, for the mentioned reason, this is not the case now. Pure monopsony is as rare as a pure monopoly. Pure monopsony is a market dominated by a single buyer. The monopsony power is the ability of a single buyer on the market to influence the price of the goods (or production resources service). Thus, the emergence of a natural monopoly is the result of economies of scale. A natural monopoly occurs when the average cost of one firm decreases as production increases and when one firm can produce all the volume of the goods it buys at lower average costs than two or more firms. Why are costs falling until the quantity needed to meet aggregate demand is produced? The answer may be high fixed costs. Local electricity, telephone, water and gas services are natural monopolies as the fixed costs of constructing electricity or telephone, water or gas pipelines are high compared to variable costs (Medema, 2007).

The legal monopoly is recognized and protected by the law in order to prevent unfair practices and give producers, authors and inventors the opportunity to exploit the fruits of their efforts. Legal monopolies are patents, copyrights, which recognize individuals' exclusive rights to receive income from certain goods and to use certain objects. Public authorities, acting in the interests of consumers or the public, often restrict the activities of producers.

The social monopoly concerns water, gas, electricity, postal and telephone services. The companies providing these services have a monopoly stemming from technical supply conditions. It would be quite impractical and even impossible to engage in such activities in a given area for more than one company, which would cause inconvenience to society and wasting of resources. Imagine how difficult it would be to call a doctor on the phone if there were two telephone companies, one for you and the other for your doctor. Or how uncomfortable it would be if dozens of gas and electricity companies had the right to excavate the streets of the city.

The effects of the monopoly are best seen in comparison with the perfect competition sector. First, the output of the monopoly at equilibrium is lower than that of the perfect competition sector. As a result, of course, the monopolized sector uses fewer resources than the perfect competition sector. Since not all economic sectors are monopolized, the monopoly undermines the optimal allocation of production factors by industry and the economy as a whole. Second, the monopoly sells them at a higher price than in the case of a perfect competition by protecting a smaller quantity of goods or services. As a result, a monopoly reduces the surplus of consumer benefits and increases the excess benefits for producers. Third, a monopoly leads to utility losses as the public benefit outweighs the marginal cost segment. It is therefore appropriate, from a public viewpoint, to produce smaller quantities of production. The monopoly, therefore, results in welfare losses or social costs caused by monopolistic power. Fourth, another example of the social costs of monopolistic power is the auction of individuals, their groups, when they spend money to bribe the government or local authorities to obtain the monopolistic rights guaranteed by the government. This is the pursuit of a rent because the economic profit of the monopoly is also called the rent – it is surplus over opportunity costs. Fifth, of course, the monopoly produces without minimum average total costs. This leads to a lack of efficiency in its activities in the long term. Meanwhile, the perfect competition sector produces at minimum average total costs (Medema, 2007).

Lack of information. Consumers need sufficient information to make informed choices based on reliable information. Many electricity companies provide little or no information on the sources of energy they supply or their emissions to the atmosphere. Since RES technologies are new compared to traditional ones, many consumers are very little aware or unaware of them. Many consumers may think that they cannot buy solar or wind energy because it is produced and supplied only when the sun shines or the wind blows. They do not know that these technologies are highly diagnostic when used in conjunction with other technologies.

Fragmented initiatives. If RES are used to supply energy to individual homes or businesses, using photovoltaic systems, fuel cells and small wind turbines, the development of RES faces additional market barriers. Building owners are aware of the most efficient locations for solar collectors; however, they have no incentive to install these facilities as they are not interested in energy savings or other benefits for tenants. Meanwhile, tenants do not have the right to modify alien buildings or have no interest in making long-term investments in buildings that do not belong to them.

4.2 Measures to Promote Renewable Energy Sources in Households

Renewable energy technologies have a significantly lower environmental impact throughout their life cycle than fossil-based technologies, in particular the fact that RES do not cause GHG emissions at the energy production stage. They can, therefore, be considered as environmentally friendly sources of energy production. Investment in RES has other advantages too such as job creation and economic growth (Lyu & Shi, 2018). Moreover, the generation of distributed renewable energy can increase the security of supply and reduce the country's dependence on energy imports.

RES play an important role in the global economy and are a key climate change mitigation measure, taking the lead in energy policy (European Commission, 2019a,b). It is therefore expected that there will be increasing interest and investment in RES technologies and their development (Chang & Shieh, 2017; Qi & Li, 2017; Zeng *et al.*, 2018; Lu *et al.*, 2020). While widely accepted globally RES have many advantages; however, their penetration in global and local energy markets is still relatively low. This problem is caused by various obstacles to the development of RES: social, economic, technological and regulatory (Painuly, 2001).

There is a wide range of research on the barriers to the development of RES and market failures that hamper the use of RES (Sovacool, 2009;Byrnes *et al.*, 2013; Huang & Li, 2013; Sovacool & Saunders, 2014; Raza *et al.*, 2015; Nasirov *et al.*, 2015; Karatayev *et al.*, 2016; Seetharaman *et al.*, 2019; Malik *et al.*, 2019) and renewable energy promotion measures (Zhang *et al.*, 2014; Boie *et al.*, 2014; Nesta *et al.*, 2014; Harrison, 2015; Polzin *et al.*, 2015; Cadoret & Padovano, 2016;Yaping *et al.*, 2016; Kilinc-Ata, 2016; Papiez *et al.*, 2018; Chen *et al.*, 2018, 2019). Nevertheless, there is little research done on the barriers to the development of RES in households, as well as policy measures to overcome these barriers in specific economic and energy sectors. It is therefore essential to group together all barriers and link them directly to policies aimed at removing these barriers, or to provide an assessment of policies and measures designed to remove these barriers in specific countries.

4.2.1   Barriers to the Use of Renewable Energy Sources in Households

The market penetration of RES is driven by a number of factors, including increasing support for RES and policies worldwide. Reducing GHG emissions is one of the main drivers for promoting RES. Increasing the share of renewables in final energy consumption allows countries to meet their commitments to climate change mitigation targets. However, some scientists are quite sceptical about the impact of climate change mitigation policies and have even developed the concept of the Green Paradox to highlight this problem (Sinn, 2015; Jensen *et al.*, 2015; Ploeg & Withagen, 2015).

Nevertheless, investors across the world gradually see RES as a profitable investment opportunity with higher income and benefits than fossil energy sources. While renewables are already successfully competing with conventional energy sources in some parts of the world, there are still a number of obstacles to the further development of RES. These barriers vary considerably depending on specific sectors and types of RES. In addition, overcoming one obstacle may reveal the existence of other important barriers and market failures discussed in sub-section 5.2 of the monograph.

As a result, the rapid market penetration and adoption of RES are hampered by numerous social, economic, technological and regulatory barriers. These obstacles have been identified by scientists as the main reasons for preventing the large-scale use of RES (Zyadin *et al.*, 2014; Nasirov, 2015).

Public opposition and unfavourable assessment of renewable energy projects are important social barriers to higher market penetration of RES. This resistance is mainly due to a lack of understanding of the advantages of RES. Other social barriers relate to the acquisition of land for renewable energy infrastructure, as land for this purpose could be successfully used for agriculture and tourism (Paravantis *et al.*, 2014; Zhao *et al.*, 2016). A large part of the utilized agricultural area, including cultivated areas, is turned into roads, building structures and other necessary infrastructure for the operation and maintenance of renewable energy plants. Other sectors such as agriculture, tourism and fisheries (Boie *et al.*, 2017) suffer as a result (Edomah *et al.*, 2017). Lack of public awareness and information barriers prevent rapidly RES from penetrating the market and competing with fossil fuels (Raza *et al.*, 2015).

In their studies, scientists have identified a very important "Not in My Backyard" (NIMBY) syndrome, which is evident when looking at renewable energy projects that face public opposition to various problems, such as negative environmental and landscape effects, and conflicts with local communities. (Nasirov *et al.*, 2015). A major problem is also the loss of other revenues from land taken for large renewable energy projects. Developing countries are usually faced with a significant shortage of skills and education about RES, and there is a lack of skilled labour for the development of RES projects. Important obstacles such as the failure to carry out the operation and maintenance work properly also cause many problems and even lead to the failure of the RES project immediately after the project implementation phase. There is, therefore, a consensus among scientists that the lack of experienced specialists and training institutes prevents RES technologies from becoming more adopted into households (Karakaya & Sriwannawit, 2015). Key policy measures to overcome social barriers to the development of RES such as awareness-raising campaigns, demonstration projects, training and capacity building were identified (Sovacool, 2009; Paravantis *et al.*, 2014; Kilinc-Ata, 2016; Seetharaman *et al.*, 2019).

The economic and financial obstacles to the development of RES are linked to the high initial capital requirements, the lack of financial institutes and investors in renewable energy projects, subsidies for fossil fuels and the consequent price competition between conventional and RES (Byrnes *et al.*, 2013); Raza *et al.*, 2015). Economic and financial barriers have so far failed to ensure the widespread use of RES in households. Thus, high initial capital costs are an important economic obstacle as RES projects require a substantial initial investment. Due to the limited efficiency of renewable energy plants and their intermittency of operation, RES projects have a long payback period (Lyu & Shi, 2018). As a result, RES projects often remain unimplemented for precisely the following reasons (Painuly, 2001). As there are fewer funding institutions providing loans and financing for RES projects, developers of RES projects face problems such as ensuring the financing of RES projects. Due to the lack of institutional experience and limited accessibility to effective risk management mechanisms, such as guarantees, RES project developers find it difficult to find appropriate financial instruments. This shows that investments in renewable energy projects are considered risky, which discourages investment in this area (Ohunakin *et al.*, 2014). Another problem is that the amount of government support for fossil energy sources exceeds the amount of RES grants, resulting in a disadvantageous competitive position for RES in the market (Byrnes *et al.*, 2013). External or intangible costs are another important economic obstacle to renewable energy projects. Total energy supply costs include research, production, distribution and consumption costs. However, the external costs related to environmental damage that are high on fossil fuels are not reflected in their cost. Due to falling fossil fuel prices, they compete with RES (Jovovic *et al.*, 2017). Researchers have identified the following economic and financial policy measures that could remove economic barriers to the dissemination of RES: subsidies and grants for renewable energy projects, green certificates and GHG emissions trading; GHG taxes or tax incentives for companies using RES; administratively set fixed electricity purchase prices for energy producers using RES and auctions of their prices, the establishment of financial institutions, preferential loans for RES projects (Zhang *et al.*, 2014; Harrison, 2015; Sun & Nie, 2015; Zeng *et al.*, 2018).

Researchers have also identified a range of technological barriers to the rapid penetration of renewable energy technologies such as lack of infrastructure, insufficient operational and maintenance capacity for RES, and lack of research and development. In addition, power plants using RES need energy storage due to the intermittent nature of RES activities (Gullberg *et al.*, 2014; Zhao *et al.*, 2016). Infrastructure bottlenecks reflect the lack of infrastructure for the integration of RES into the energy supply system, as there are a number of problems related to system flexibility and the limited capacity of the electricity grid to use energy from RES (Boie *et al.*, 2014; Raza *et al.*, 2015). It should be stressed that the costs of developing the infrastructure necessary for RES, including transmission lines for connection to the network, are very high. Moreover, due to complex technologies and established procedures and guidelines, including reliability and other standards, it is not possible to use RES technologies widely (Nasirov *et al.* 2015). The lack of research and development prevents RES technologies from competing effectively with conventional power generation technologies using fossil fuels. Moreover, the high risk associated with RES technologies already mentioned prevents businesses and governments from investing sufficiently in R&D activities (Cho *et al.*, 2013). Furthermore, there is no culture of exploitation and maintenance, which is also an important obstacle to the development of RES; as these technologies are new and are still under development the expertise is very limited. The following key policy measures have been identified to address the technological barriers to RES development: the State support for the supply of renewable energy infrastructure such as energy storage, financing, equipment and parts needed for the operation of renewable energy technologies, the abolition of import taxes, VAT and other tax reliefs (Boie *et al.*, 2014; Edomah *et al.*, 2017; Lyu & Shi, 2018).

Regulatory barriers also play an important role in halting the rapid development of RES. Renewables' penetration into the market requires strong political support and a well-developed regulatory framework supporting and promoting renewable energy. However, in many developing countries there is strong political opposition to renewable energy projects such as institutional corruption and lobbying of fossil fuels or nuclear energy (Ohunakin *et al.*, 2014). Researchers have identified the following regulatory and institutional barriers to the development of RES: the absence of national policies to support renewable energy, red tape and overregulation; insufficient incentives; unrealized government targets in the field of renewable energy; lack of standards and certificates (Stokes, 2013; Sun & Nie, 2015). The removal of barriers to RES regulation requires a strong regulatory policy for the energy industry. In the absence of effective policies, the various market players, government institutions and departments do not have a clear understanding of the implementation of the necessary support measures. However, the main problems relate to unpredictable energy policy and insufficient political confidence and insufficient support for RES development projects (Zhang *et al.*, 2014). There are other regulatory barriers such as the lack of clearly allocated responsibilities; complex authorization procedures; other problems with permits and land acquisition; limited planning guidelines, etc.

Insufficient fiscal incentives are also an important problem for the development of RES (Sun & Nie, 2015). Other obstacles include information asymmetry and lack of information, divided initiatives, lack of transparency, subsidies for conventional fuels and inability to integrate external energy production costs into the energy price (Browne *et al.*, 2015). Bureaucratic procedures for the deployment of RES are considered to be the biggest barrier to investment in renewable energy projects (Huang *et al.*, 2013). There are other regulatory barriers such as ineffective policy-making or inconsistent policies, unclear feed-in contracts and other agreements. It should be stressed that the various barriers to the development of RES often overlap.

Although many countries have set renewable energy targets in their long-term strategic planning documents, there is a clear gap between the policy objectives and the real results achieved (Malik *et al.*, 2019). Policy makers lack awareness of the real objectives and there are many gaps in the implementation process itself. A prepared policy should provide a clear understanding of the necessary legislation so that business can have confidence in future RES promotion policies. There are a number of policy measures aimed at removing regulatory barriers: the share of RES in final energy, electricity and heat energy, transport and other commitments and powers; renewable electricity and renewable energy procurement for transport; renewable energy quotas and commitments. They should be based on other effective policies such as renewable energy standards, building codes and standards, specific technology mandates (Boie *et al.*, 2014; Cadoret & Padovano, 2016; Kilinc-Ata, 2016; Papiez *et al.*, 2018; Malik *et al.*, 2019).

Table 4.1 shows the barriers to market penetration of RES and measures to overcome these barriers.

As can be seen from Table 4.1, the social, economic, technological and regulatory barriers to the development of RES require targeted policies to

TABLE 4.1

Barriers to market penetration of renewable energy sources

	Barriers	Description	Policy measures
Social barriers	Barriers to public awareness and information	Insufficient information on the benefits of all types of RES; insufficient awareness of RES technologies; insufficient information on the environmental impact of RES technologies.	Information dissemination and education measures: information dissemination campaigns; provision of information in social media; social marketing; training and other awareness-raising measures in schools, universities, at the workplace, etc. Standards and certification and increasing consumer confidence.
	"Not in My Backyard" (NIMBY) syndrome	NIMBY syndrome means that people support RES in general, but not in their neighbourhood, and renewable energy projects are therefore confronted with individual citizens' resistance to the NIMBY syndrome.	Demonstration or pilot projects relevant to verifying the suitability of the site. Greater stakeholder satisfaction should be achieved by raising awareness of the benefits of RES, e.g. by creating more jobs, reducing environmental damage and mitigating climate change. Voluntary measures, corporate social responsibility initiatives.
	Loss of income	Large-scale RES projects require large-scale landscape areas, which is why RES projects require a lot of land. In order to achieve this, the loss of income of farmland, fishing and tourism businesses is often affected.	Dissemination and education measures: information dissemination campaigns; provision of information in social media; social marketing about the positive impact of RES and their environmental benefits.
	Lack of experienced specialists	The lack of technical specialists (designers, financiers, construction, operation and maintenance specialists) and training institutes hampers the market penetration of RES technologies.	Capacity building measures such as the integration of RES technology courses into higher education and vocational training programmes.

Economic barriers	High initial capital investments	Investors lack capital investment and the long payback period for RES projects is an additional barrier to the high costs of initial capital.	State agencies supporting RES projects from the structural and other funds are established. Financial support instruments such as grants, subsidies to RES projects, soft loans, loan guarantees, etc.
	Lack of funding institutions	There are fewer financial agencies to finance renewable energy projects; therefore investments in renewable energy projects are considered to be riskier and motivating investors in renewable energy technologies.	
	Subsidies for fossil fuels exceed subsidies for RES	Government support for fossil fuels, exceeding the support for RES, hampers the development of RES technologies.	The abolition of environmentally harmful subsidies (subsidies for fossil fuels).
	Non-integrated external costs in the production of fossil fuels	The external costs of producing energy from fossil fuels due to negative health and environmental impacts are not included in the energy price, making energy produced from fossil fuels cheaper and more competitive.	Policies and measures to integrate the external costs of energy production, i.e. by increasing environmental taxes on fossil fuel production technologies, subsidies for renewable energy, tax incentives and RES credits.
	Falling fossil fuel prices increase their competitive advantage	The current downward trend in fossil fuel prices has a negative impact on the market penetration of renewables.	
Technological barriers	Limited infrastructure for the development of RES	RES-based plants are usually located in remote locations where additional transmission lines are needed to integrate RES generators into the network, as well as the need to upgrade networks which would have an impact on the cost of RES projects.	Government support for RES infrastructure. Implementation of alternative infrastructure such as energy storage devices, charging infrastructure, the import tax on RES technologies, VAT, etc.
	Technology complexity is mainly related to intermittency and energy storage requirements	There is a lack of standards, recommendations and practices for working with RES technologies, which are characterized by intermittent work. The main technical problem facing RES technologies today is the need for energy storage, which also has an impact on the increase in energy generation costs. The shortage of RES equipment, components and spare parts significantly increases production costs.	

(continued)

TABLE 4.1 (Continued)

Barriers to market penetration of renewable energy sources

	Barriers	Description	Policy measures
	Lack of investment and opportunities for R&D in renewable energy technologies	Investment in R&D of RES technologies is insufficient to ensure the competitiveness of RES compared to conventional technologies. RES technologies have a higher risk profile and require additional R&D capacities.	Funding for R&D, demonstration and other financial risk mitigation measures.
Regulatory barriers	Ineffective government measures	Unstable energy policy, lack of effective policies for integrating renewable energy and government agencies, lack of knowledge, resources and expertise.	A strong, long-term, stable and effective regulatory policy, including financial measures such as feed-in payments, RES certificates, subsidies and RES projects, soft loans, is needed to increase the competitiveness of RES; loan guarantees, etc. There is also a need for a policy that offers a clear understanding of the important regulatory issues.
	Insufficient financial initiatives	Lack of financial incentives leads to high costs, which hinders the development of the RES industry.	
	Administrative and bureaucratic complexity	Due to the lack of coordination between different agencies and bureaucratic barriers, the development of the RES project requires a long time, and RES project developers face higher costs in obtaining authorizations and licences.	Reduced bureaucratic barriers by giving priority to grid-connected power plants using RES.
	Unjustifiable government commitments	There is a huge gap between the political objectives set by governments and the real results.	Obligations and other powers for the share of RES in fuels; procurement of renewable electricity, RES-using vehicles; RES quotas and commitments. Established general or specific technology/fuel targets supported by other effective policies such as RES portfolio standards, building codes, special technology powers.
	No standards and certification procedures	In the absence of standards and certification procedures to ensure that equipment produced in other countries is manufactured in accordance with the standards desired by the importers.	Vehicle standards; low carbon standards, vehicle emission standards, etc.

Source: Created by authors based on Streimikiene *et al.* (2020).

effectively remove these barriers. However, the success and impact of this policy on overcoming barriers to RES need to be assessed on the basis of concrete examples. The following sub-section assesses the strengths and weaknesses of measures to promote the use of RES in households in the EU Member States.

4.2.2 Assessment of Measures for Promotion of Renewable Energy Sources in Households

All measures to promote RES may be divided into the following main groups: setting targets and strategic obligations for RES development, financial and fiscal instruments; market instruments; RES portfolio standards and obligations; building codes and standards; information measures; demonstration projects, research and development; public procurement, alternative fuel or recharging infrastructure and voluntary schemes and initiatives.

The strengths and weaknesses of the policy measures used to overcome barriers to the use of RES in the household sector are shown in Table 4.2.

As can be seen from the information provided in Table 4.2, the main measures to overcome obstacles to the development of RES in households are envisaged in the electricity, heating and cooling sectors as well as in the transport sector. Tax incentives for biofuels are the main means of promoting the development of RES in the transport sector. A vehicle tax was introduced in several EU Member States in Latvia in 2018. An operating fee is set for vehicles equipped with an internal combustion engine first registered after 31 December 2008. The vehicle operating fee depends on the vehicle's emissions in grams of carbon dioxide (g) per kilometre (km).

Almost all EU countries have fixed feed-in prices for electricity from RES, while some countries, such as Latvia, have fixed feed-in prices for electricity from combined cycle power plants in the heating and cooling sector. In all EU countries, subsidies for RES projects prevail among financial measures to promote RES in order to reduce the burden of initial large capital investments on developers of RES projects. These subsidies are mostly financed through the EU structural funds and allocated to RES projects, determined for a specific programming period. In addition, some EU countries have a net electricity accounting system and a detailed invoicing method.

All EU Member States have set RES development goals in their strategic planning documents, as this is required by EU directives. In general, the electricity sector is at the heart of the policy measures to promote the development of RES. However, there are other important sectors that face significant social, technological and regulatory barriers such as the transport sector. Climate change management laws and strategies are essential for the achievement of renewable energy targets in all EU countries, as the EU's energy and climate change policies are integrated and the RES development goals for 2020, 2030 and 2050 include RES, energy efficiency and reduction of GHG emissions.

TABLE 4.2

Strengths and weaknesses of policy instruments to overcome barriers to renewable energy sources

Policy measures	Barriers	Strengths	Weaknesses
Goals: the general or specific objectives of the technology or fuel; strategic planning of RES.	Regulatory barriers: impractical government commitments; unstable and ineffective policies.	Demonstrate clear directions for strategic planning; sends clear signals to consumers and industry.	These measures are not effective; additional strategies and measures are needed to achieve the goals.
Financial measures: various subsidies and grants, tax incentives and investment subsidies.	Economic barriers: high initial capital costs; less funding institutions; increased subsidies for fossil fuels.	Improve the competitiveness of renewables compared to fossil fuels and can significantly contribute to overcoming the higher capital costs of renewables.	The level of support is changing for political reasons and changing government priorities. Sometimes they do not provide the right signals to investors.
From an administrative point of view, pricing measures have been identified as follows: fixed electricity purchase prices.	Economic barriers: high initial capital costs; instead, there are no integrated external costs.	Provides support for a long time.	May cause high costs of snowballs; does not address significant capital costs in advance.
Taxes on coal, energy or fuel: the tax on GHG emissions per km, etc.	Non-integrated external costs; falling fossil fuel prices.	Provide an important price signal and integrate external effects. Over time, policies can be changed by raising taxes.	Fees are difficult to implement for political reasons; exemptions, often granted to certain industries and rendering taxes ineffective.
Certificates on sale: green, white or emission allowance certificates.	Non-integrated external costs; falling fossil fuel prices.	Market mechanisms providing additional income for RES generators.	Require a compliance and enforcement mechanism, high implementation costs.
Competitive pricing tools: auctions.	Non-integrated external costs; falling fossil fuel prices.	Design flexibility and real-price option.	There is a high risk of expelling small or new players from the market.
RES portfolio; obligations and powers: the share of RES in fuels used in transport; RES heat quota; requirements for the installation of solar water heaters; obligations of zero-pollution vehicles.	Regulatory barriers: lack of standards and certification; administrative and bureaucratic complexity.	Provide certainty to the levels of deployment; sends clear signals to the industry.	The heat sector is less efficient than the electricity sector. In most cases, it applies only to a very low proportion of heat demand in new buildings.

Construction standards: energy performance requirements for energy efficiency or RES.	Social barriers: barriers to public awareness and information.	Addressing important social barriers; provides an opportunity to combine energy efficiency with renewable heat demand.	In most cases, the standards are applied only to new buildings, thus meeting a small part of the heat demand; rarely applicable to existing buildings.
Prohibitions on the use of fossil fuels for heating.	Social barriers: barriers to public awareness and information.	Mandatory are more effective than voluntary.	Suitable alternatives must be available to replace fossil fuels with RES.
Information tools: information campaigns and energy labelling of buildings and installations.	Social barriers: barriers to public awareness and information; the NIMBY syndrome.	This is necessary to ensure awareness of the potential, costs and benefits of the RES use, including their external benefits.	Most effective when implemented as part of customized energy advice that is very expensive to organize and implement effectively.
Standards and certification: standards for low-carbon fuels: vehicle emission standards; vehicle standards.	Regulatory barriers: lack of standards and certification.	Important for supply chains and increase consumer confidence.	Without financial incentives, the impact is not likely to be significant.
Continuous provision of information on energy consumption through counters or billing.	Social barriers: barriers to public awareness and information.	Can make energy savings for end-users and the system as a whole, which would help to reduce transmission, distribution losses at peak times.	Can create cross-subsidization for those consumers who consume the energy they produce. There is a risk that retail tariffs do not accurately disclose actual electricity over a given period of time.
Capacity building: installers' training.	Social barriers: lack of experienced specialists.	Important for supply chains.	It is doubtful whether they can independently influence the development of RES. Additional policies and measures are needed.
Demonstration projects.	Social barriers: barriers to public awareness and information; the NIMBY syndrome; loss of alternative income; technological barriers: limited RES infrastructure; the complexity of the technology associated with intermittency.	It is important to check the local suitability of the RES technology and to combat the NIMBY syndrome.	It is doubtful whether they can independently influence the development of RES. Additional policies and measures are needed.

(continued)

TABLE 4.2 (Continued)

Strengths and weaknesses of policy instruments to overcome barriers to renewable energy sources

Policy measures	Barriers	Strengths	Weaknesses
Funding for research and development and demonstration: for example, capital-intensive advanced RES, biofuels, etc.	Technological barriers: there is a lack of investment and opportunities in research and development and RES technologies.	It is necessary to ensure the launch of commercial pilot projects involving technologies with a long-term perspective for market expansion but with a high risk of investment.	The financial risk associated with the potential failure of the project.
Renewable heating and renewable energy procurement for transport, etc.	Regulatory barriers: inadequate fiscal initiatives.	Can be used as a starting point for increasing the deployment of RES technologies in general.	This is only possible by ensuring a certain share of demand and this measure needs to be complemented by other measures to facilitate the market penetration of RES technologies.
Alternative fuel or recharging infrastructure is installed.	Technological barriers: limited RES infrastructure; the complexity of the technology associated with intermittency.	The need for greater development of alternative fuel vehicles.	Infrastructure costs need to be balanced, as demand is low in the early stages of the RES development.
Voluntary programmes	Social barriers: barriers to public awareness and information; the NIMBY syndrome; regulatory barriers: ineffective policies; impractical government commitments.	Empower voluntary initiatives by saving public and consumer costs.	It requires awareness-raising programmes. Their implementation does not necessarily take into account the objectives set by the government for RES.

Source: Created by authors based on Streimikiene *et al.* (2020).

There are some important differences between the policies implemented by EU countries to overcome barriers to market penetration of RES. Some countries, such as Estonia and Latvia, had systems in place to promote electricity produced from RES. However, Latvia changed the Green Certificate system, which was considered to be a poorly balanced support measure, although it allowed RES producers to earn some income during certain periods, although in certain periods it caused particular problems. The unclear system of this type does not support the continuous development of RES; therefore, Latvia has initiated a switch to an auction where consumers could win due to cheaper energy prices and RES; renewable energy producers by participating in auctions and offering the best price can secure the necessary cash flows for themselves. This, in turn, allows them in the future to offer cheaper electricity produced from RES to the electricity market.

In Estonia, environmental (CO_2) taxes on fuel combustion are applied additionally. This tax was introduced in the country taking into account the dominance of shale oil in the district heating and electricity sector. There are no such taxes in the other Baltic States.

4.3   Energy-Saving Promotion Measures in Households

Although most countries in the world have implemented policies to promote the renovation of buildings, this policy did not actually reduce energy consumption and GHG emissions (Chang & Shieh, 2017; Zundel & STIER, 2011; Janda & Parag, 2013; Janda *et al.*, 2014). Most countries have implemented financial incentives such as capital subsidies, grants, subsidized loans and discounts to encourage building owners and residents to invest in energy efficiency improvement measures and equipment, but multi-apartment buildings may face significant barriers to large-scale energy renewal (Scott, 1977; Risch, 2012; Crilly *et al.*, 2012; Burger, 2013; Killip, 2013; Sirombo *et al.*, 2017).

Therefore, due to the lack of understanding or lack of adequate and reliable information, public subsidies and loans and other financial measures cannot be used by building owners or housing associations (Golove & Eto, 1996; Janssen, 2004; Assadi *et al.*, 2012; Friege & Chappin, 2014; Claudy & O'Driscoll, 2008). In addition, potential large-scale energy renovators have difficulty understanding the efficiency and benefits of energy efficiency measures. There are also financial difficulties such as difficulties in accessing capital that also hinder the investment in large-scale energy renewal (Organ *et al.*, 2013; Friege & Chappin, 2014; Horne & Dalton, 2014). Even if apartment owners have access to capital, they cannot invest in the energy renewal of residential buildings due to the high risk of such an investment. Risks are considered to be related to limited rationality and uncertainty about the benefits of large-scale residential renovation due to unpredictability in future energy prices (Kahneman & Tversky, 1979; Wilson & Dowlatabadi, 2007; Banfi *et al.*, 2008).

4.3.1 Barriers to Energy Efficiency Improvement in Households

Researchers have discussed extensively on defining and grouping barriers to energy savings in households (Golove & Eto, 1996; Asadi *et al.*, 2012). The main barriers to energy efficiency improvement are summarized in Table 4.3. Table 4.3 shows that there are six main barriers to improving energy efficiency in households: economic or financial barriers, hidden costs, market failures, behavioural and organizational barriers, information barriers and political and structural barriers.

The following sub-section discusses barriers to energy efficiency improvement in households, as well as an assessment of the policy instruments to energy efficiency promotion according to their ability to overcome these barriers and realize the potential for energy savings and reducing GHG emissions in residential buildings.

4.3.2 Assessment of Energy Efficiency Promotion Measures

Scientific literature (Sirombo *et al.*, 2017; Killip, 2013; Risch, 2012; Crilly *et al.*, 2012) argues that already implemented policy schemes do not sufficiently remove barriers to decision-making by house owners. The traditional energy renewal policy focused the most on financial incentives such as preferential loans and subsidies. In addition to financial incentives, some researchers stressed the need to implement other policy schemes, such as regulatory and information measures, and to ensure greater involvement of market players by increasing the participation of experts in energy efficiency and renewal (Janda & Parag, 2013; Janda *et al.*, 2014).

All energy efficiency support measures to overcome energy efficiency barriers in households are divided into regulatory and control measures, measures based on economic and market mechanisms, fiscal instruments and initiatives, as well as the dissemination of information and voluntary measures.

First of all, the detail regulatory and control measures which include energy class marking of buildings, electrical equipment standards, energy efficiency commitments and quotas, mandatory energy audits, equipment marking and certification programmes and demand management programmes will be critically discussed (Streimikiene *et al.*, 2020).

Mandatory building standards for energy consumption and energy class marking or certification of buildings have been introduced in almost all developed countries in the world that include these energy efficiency

TABLE 4.3

Barriers to energy efficiency improvement in households

Main barriers	Origin of barriers	Examples of barriers
Economic and financial barriers	High investment cost of energy savings due to the low internalization of external energy production costs	High initial costs related to energy refurbishment of multi-apartment buildings combined with low incomes and insufficient capital.
Hidden costs	Expenditure not directly included in the financial flows of energy renewal	Costs due to high operational risks, high transaction costs, etc.
Market failures	Market constraints that do not compensate for energy-saving costs and energy-saving benefits	Fragmented market structure; a dilemma of the lessor/tenant, sharing incentives; administrative and regulatory barriers; information asymmetry.
Behavioural and organizational barriers	Behavioural models for individuals and undertakings that impede the implementation of energy efficiency technologies and practices	A tendency to ignore low energy saving potential; difficulties in reaching a joint decision by apartment owners on large-scale renewal due to conflicting interests; non-payment and theft of electricity; traditions, behavioural patterns and lifestyles, etc.
Barriers to information	Missing information on energy-saving potential	Lack of knowledge of apartment owners, building managers, construction companies, politicians and others.
Political and structural barriers	Political barriers, economic and energy system complicates investment decisions for energy renewal	The process of drafting local legislation is slow; huge disparities between regions at different economic levels; insufficient enforcement of standards; lack of detailed guidance and experts; lack of incentives for investment; lack of political interest; lack of certification services for equipment; the inadequate level of energy services, etc.

Source: Created by authors based on Streimikiene *et al.* (2020).

standards, while many developing countries now legislate on the development and implementation of such codes. In many cases, these building standards and energy class markings are designed to regulate new buildings, but recently many governments in developed countries have amended their laws to cover the renovation and refurbishment of existing buildings. Most standards are based on building performance: that is, they determine maximum thermal conductivity through building envelopes and heating/cooling demand, as well as require home infrastructure equipment such as heating and air conditioning systems, ventilation, water heaters and even pumps and lifts to meet certain set energy performance standards. Standards for equipment or materials may also be used in buildings. Energy standards and energy labelling requirements for buildings are almost always more successful when they are mandatory rather than voluntary. Mandatory building codes can help boost energy efficiency investments and reduce barriers to information limitations and asymmetry (Streimikiene *et al.*, 2020).

When a building is put into service, a systematic testing process is necessary to ensure that the building systems have been designed, installed and put into service in accordance with the design requirements and the building owner's operational needs. As regular service increases the life of a car, the proper start-up of the energy systems of buildings is essential for the efficient operation of the building throughout its life cycle. Studies in the US have shown that the proper commissioning of buildings has produced impressive results: energy savings amount to 38% from cooling and 62% from heating, while the total energy savings account for more than 30% (Levine *et al.*, 2007).

Developed and many developing countries have introduced standards for equipment for energy-using products, such as lighting, heating and cooling equipment and personal computers, and widely apply marking and certification systems for these devices. For example, the Top Runner programme introduced in Japan requires all new products to meet the required efficiency levels by the specified date (Geller *et al.*, 2006). For example, there are strict lighting requirements like fluorescent lamps and LED bulbs are 75% more efficient than conventional incandescent bulbs. Significant progress has also been made in improving the efficiency of air conditioners, ventilation systems and other appliances. Experience in many countries shows that device standards are cost-effective and very effective in reducing transaction costs for buyers and manufacturers (Stadelmann, 2017; Gillingham *et al.*, 2004; UNEP SBCI, 2009). The standards of devices are most effective when used in conjunction with the marking programme for devices. However, one of the drawbacks of standards is that they do not encourage innovation beyond the target and therefore require periodic updates. In countries that import all electrical appliances, local testing and certification are very expensive as they require capital investment in testing equipment and training. It would be useful to have an international testing and certification system so that laboratory tests carried out in one country can be used in another. Some governments have also tried to implement standards for devices, including fiscal measures. The most popular measure is a reduced import tax or VAT on energy-efficient products.

Another important regulatory and control instrument is a compulsory energy audit. It constitutes a consistent extension of building standards and the procedures for the commissioning of buildings. In many European and other countries, governments have made energy audits mandatory for public buildings as well as for other important energy-using sectors such as specific industrial and large commercial users. In the EU, these audits also determine the proper maintenance of energy-using equipment, as for example boilers.

Although comprehensive energy audits are quite expensive and require a high level of technical skills, it has a great advantage over other strategies, as it provides practical data and can serve a large number of customers in a short period of time. In particular, in order to implement all the recommendations of the energy audit, in developing countries, more attention needs to be paid to improving the quality of auditors by training and providing practical and financial support to owners and managers of the buildings audited (Streimikiene *et al.*, 2020).

Energy efficiency obligations (EEOs) and quotas are a legal obligation for electricity and gas suppliers and distributors to save energy at their customers' premises and, in turn, increase energy efficiency. The savings targets are set by the government or, in the case of Ireland, by the regulatory body. The measure was adopted in 2007 in the UK, Flanders, Denmark, Italy, France and Ireland. In all countries that introduced EEOs, the residential sector was the target customer of these suppliers. Energy suppliers who fail to meet the government's objective are fined. In many cases, energy suppliers can purchase the difference between savings and targeted savings through the white certificate scheme. Energy efficiency obligations have several advantages: they can be administered cheaply; they do not need to be included in the State costs; and despite the fact that energy prices generally rise from 1% to 2%, this measure helps to avoid regression social impacts on low-income consumers (UNEP SBCI, 2009).

Energy demand management programmes are mandatory or voluntary planning and implementation of energy efficiency programmes by utilities. Such activities include consultancy services for individual consumers (e.g. advice on new heat pumps or electrical installations, energy audits), awareness campaigns aimed at changing energy consumption behaviour or promoting new appliances (e.g. meters and low-energy light bulbs), and information bills for electricity (Hein Nybroe, 2001). In the past, these programmes were usually initiated as a means of relieving pressure when energy demand exceeds generation capacity, and they have often been discontinued due to changes in market conditions or regulatory environments. The EU is facing a decline in voluntary energy demand management programmes, but the liberalization of electricity markets has also enabled new policy initiatives in this area (Palmer, 1999; Eyre, 1998; Evander *et al.*, 2004). Energy demand management programmes have contributed to market transformation by encouraging local manufacturers, importers and distributors to consider the production and import of more efficient appliances and encouraging consumers to buy these new products (Brulez *et al.*, 1998).

The regulatory and control measures are closely related to economic and market mechanisms such as energy-saving contracts, collective procurement, white energy efficiency certificates and Kyoto flexible mechanisms, and the combination of these measures allows for the best results (UNEP SBCI, 2009).

Some countries have recently linked their energy efficiency obligations to sales of energy-saving certificates. Energy efficiency certificates or "white certificates" are a relatively new policy instrument first introduced in New South Wales, Australia in 2003, followed by Italy in 2005 and France in 2006. White certificates are issued by independent certification bodies confirming the outcome of final energy consumption efficiency measures (Bertoldi & Rezessy, 2006; Capozza, 2006). Certificates are used to facilitate energy savings under energy efficiency obligation schemes, although to date white certificates have been marketed by companies directly involved in those schemes. While white certificate programmes may promote energy savings, their effectiveness depends on rigour in setting energy efficiency requirements and on inspection and enforcement systems (Ries *et al.*, 2009).

An energy performance contract means that a contractor, usually an energy service company (ESCO), guarantees a certain amount of energy savings in a given location over a specified period of time, implements appropriate energy efficiency improvements, and this service is paid from the estimated reduction in energy consumption resulting from energy savings (EFA, 2002). This measure is becoming increasingly popular for the implementation and financing of energy efficiency projects in buildings, as no public expenditure or market intervention is needed to exploit the potential of cost-effective energy savings. However, for the ESCO activities to be effective and popular, a number of conditions are needed such as the development of financial markets to borrow funds for energy efficiency projects, non-subsidized energy prices and a favourable legal, financial and business environment in the country. To date, ESCO has been shown to be effective in Germany, the US and Hungary as well as in China and Brazil; however, in other countries, such as India, they have been less successful (Urge-Vorsatz *et al.*, 2007). For many developing countries, ESCO projects are funded by bilateral and multi-lateral supporters.

Collective technology procurement is another example of how consumers can pool their market power. It is a voluntary measure used in both the public and private sectors, where customers purchasing large amounts of energy-using appliances and equipment cooperate in order to influence the market for more efficient products. Their requirements generally include energy effi-ciency specifications that are in line with or even exceed, global best practice (EFA, 2002). This measure makes it possible to increase the commercial avail-ability of new technologies to all buyers. Public procurement regulations are also a very powerful measure to replace appliance markets in favour of devices with higher efficiency (UNEP SBCI, 2009).

Kyoto flexible mechanisms, such as Clean Development Mechanisms and Joint Implementation projects, have been very limited to improving energy efficiency in buildings, although there were options to implement them and obtain GHG emission reduction credits in developing countries through funding for climate change mitigation projects related to energy savings in buildings.

Many governments use fiscal instruments such as capital subsidies, grants, subsidized loans and public benefit taxes to encourage building owners and residents to invest in energy efficiency measures. In particular, governments have directed their support to the heating and cooling sector due to high energy consumption in buildings, in particular, due to poor thermal insulation. Financial incentives are most often used to promote the insulation and modernization of external walls, ceilings, attic, floor, window frames and connections, as well as water heaters and storage tanks, boilers and water pipes (UNEP SBCI, 2009).

Capital subsidies are very common in the residential sector in order to overcome the financial obstacles associated with large capital investments (WEC, 2008). These subsidies were used to finance better insulation of buildings such as roof insulation in the UK, more efficient equipment such as refrigerators in Germany and energy audits in France. The German and Slovenian subsidy schemes were very effective (WEC, 2004). Limiting subsidies either to a short period of time to facilitate the introduction of new technologies in the market or to a specific target group in need enhances the effectiveness of the measure (Jeeninga & Uyterlinde, 2000). Some governments have also put in place preferential loan schemes under which loans for the installation of energy efficiency equipment are granted through interest rate subsidies. Green or energy-saving mortgages (EEMs) are often referred to as "green mortgages" and are loans that give the borrower a lower interest rate or a higher loan than is normally allowed. They apply to energy-saving action to deliver housing or to purchase a house that meets certain energy efficiency standards. The economic basis of eco-friendly mortgages is that energy-saving houses will save money for house owners; therefore, higher income allows the beneficiary or the house owner to borrow more. It is still difficult to determine whether green mortgages will become the norm in the future, as their attractiveness is partly dependent on the value of savings and therefore on energy prices for both lenders and borrowers. Besides, they are also a good example of how the financial sector can take the risk of energy savings if a reliable verification system is known. Therefore, one of the necessary conditions for introducing "green mortgages" is the existence of nationally recognized energy performance standards (Streimikiene *et al.*, 2020).

Some governments prefer fiscal measures, such as *tax incentives*, to encourage investment in energy savings and improve efficiency in buildings. In the residential sector, tax credits are the most popular, while the commercial sector uses tax discounts and accelerated depreciation. Almost 40% of the OECD countries offer tax credits for energy efficiency measures. The World Energy Council found that fiscal incentives are considered to be better than subsidies because they are less costly; however fiscal incentives usually have a poor performance in an economy in recession or transition (WEC, 2008).

Public utility charges are a new mechanism defined as the collection of funds from the functioning of the energy market, which can then be focused on energy demand management programmes and other energy efficiency activities (Crossley *et al.*, 2000). They are therefore in nature similar to an energy tax whose revenues are usually partly or fully invested in energy efficiency. Public utility charges can raise funds for energy efficiency improvement measures and possibly accelerate market transformation. However, their effectiveness in terms of overall GHG savings is low: studies in the US showed that 0.4% of all electricity sold was saved (Kushler *et al.*, 2004).

Energy and CO$_2$ taxes are a popular fiscal measure in EU countries. While the relationship between energy demand and energy prices is complex, energy prices are a key factor in determining the attitude and behaviour of energy consumers. Some countries have developed direct taxes on domestic fuel or carbon emissions as a means of promoting energy savings. In most cases, the tax is paid by the final consumer (e.g. households), but the tax can be collected anywhere in the supply chain (Crossley *et al.*, 2000). Several European countries levy tax on energy consumption or energy-related CO$_2$ emissions. Taxes have a number of advantages: they have a direct impact on the entire life cycle of buildings and can reinforce the impact of other measures, such as standards and subsidies, or allow more profitable investments in energy efficiency. Energy or CO$_2$ taxes are also a beneficial measure to secure funding for other energy efficiency programmes such as discounts for energy efficiency programmes, loans or special assistance to low-income households to improve their energy efficiency. For example, countries can use such taxes to create an energy efficiency investment fund that finances initial energy efficiency investments in buildings that meet a minimum energy consumption benchmark for a particular type of building in the country. The effectiveness of energy taxes is also disputed, in particular because they may have serious social and political consequences (Crossley *et al.* 2000). Similarly, governments did not want to abolish energy subsidies, which, like taxes, would raise the price of energy. In order to mitigate negative social impacts, governments can use the money saved or collected under this policy to create other energy-saving mechanisms (Streimikiene *et al.*, 2020).

Information and voluntary measures include voluntary and negotiated agreements, public leadership programmes, including public procurement regulations, education and information programmes and detailed invoices, and the disclosure of details of energy use information.

Voluntary or negotiated obligations include obligations of energy companies or construction organizations regarding the implementation of energy efficiency improvement measures in buildings, but these programmes are more effective when combined with regulatory and financial incentives (UNEP SBCI, 2009).

Education and information programmes encouraging consumers to save energy and implement energy efficiency measures are most effective when combined with other policy measures. For example, the efficiency of electrical equipment standards increases when mandatory labelling programmes are presented together, as shown by the results in China (Lin, 2002). Mandatory and voluntary labelling schemes are used in many countries around the world, including many developing countries. More than half of the Asian countries and 90% of the South American countries have labelling programmes. The US "Energy Star" programme is an example of a successful voluntary labelling scheme (Gillingham *et al.*, 2004). In general, mandatory labelling is more effective than voluntary labelling, as it avoids the problems of inefficient devices that are not labelled but are cheaper. There are also effective disclosure programmes to encourage energy consumers to reduce consumption. Studies have shown that displaying the energy consumption of appliances is more effective than billing energy consumption data. This is one of the reasons why governments in some countries encourage the private sector to install energy meters for appliances in new buildings. Awareness-raising campaigns are also widespread in many countries, including programmes providing advice to energy savings, consultations, feedback and assessment of energy savings, primary school programmes and public awareness-raising campaigns to promote energy savings. These programmes are more effective for households, but not for the commercial sector. In Brazil, for example, the effectiveness of information programmes exceeded the effectiveness of many other policies (Dias *et al.*, 2004). Information dissemination programmes are particularly important in developing countries where the lack of information is seen as a major barrier to energy investments in energy efficiency and RES (Evander *et al.*, 2004).

The disclosure of information on energy use (e.g. annual energy consumption in kWh per building area or per capita) and the disclosure of emissions (e.g. annual GHG emissions per building area or per capita) or the inclusion in utility bills support mandatory and voluntary audits and provide information to users of buildings on how to reduce energy consumption and GHG emissions. However, consumers still need to take concrete action on the basis of this information. For this reason, such measures are most useful in combining them with other strategies such as subsidies for recommended energy efficiency measures, training and information measures or fiscal incentives to replace old energy equipment in buildings. The audited buildings may also receive awards or certificates, thereby increasing public recognition for the successful implementation of energy efficiency measures (UNEP SBCI, 2009).

Public procurement is an important tool in many parts of the world, which can lead to the wider use of energy-efficient equipment, materials and buildings, by following the requirements for public procurement of these facilities or related services.

Equipment using renewable energy in buildings and other installations faces the same obstacles as energy-saving installations. In order to overcome these barriers, some countries have simply opted for regulations that make renewable energy installations, such as solar collectors for water heating, to be mandatory for new buildings; however, so far, the most widespread policy measures to promote the use of RES in buildings are the use of subsidies, grants and fiscal incentives. For example, Japan has increased national solar subsidies for schools, hospitals and railway stations from 33% to 50%, as well as provides such subsidies to households. Ireland, Germany and Luxembourg provide subsidies or grants to install solar water (and sometimes space) heaters in residential, public and commercial buildings. "Eskom", the South African utility company, has also recently started a solar hot water subsidy programme that provides USD 200–350 per household. It is also worth noting that in developing countries, particularly those with relatively low electrification rates, there has been tremendous progress in disseminating renewable energy appliances, sometimes in conjunction with rural electrification programmes. The electrification programme for Indian villages included over 435,000 home lighting systems and 7,000 solar-power water pumps (UNEP SBCI, 2009).

Renewable energy technologies are becoming increasingly accessible and flexible; therefore, the interest in both new and existing buildings is increasing. The design of a "green building" or "sustainable building" is done to combine design and technologies, mainly renewable energy systems, in order to reduce GHG emissions. For example, passive houses are houses that maintain a comfortable internal climate without active heating and cooling systems. Their additional energy requirements can be fully met by the use of RES. At the same time, zero-energy buildings are buildings where energy provided by on-site RES is equal to the energy used by the building. In addition, energy can be stored on-site using various energy storage systems. Zero-energy buildings are usually connected to the main electricity grid in order to meet possible fluctuations in demand, especially as some buildings will generate more summer energy and consume more in winter. Several exciting model projects have been built over the past few years, including a zero-energy housing project in the Netherlands and the headquarters of the Malaysian Energy Centre in Kuala Lumpur (UNEP SBCI, 2009). But perhaps the most interesting projects taking place today are energy-plus buildings – buildings that produce more energy than they consume over a year. The extra energy is usually electricity, produced with solar cells, solar heating and cooling, insulation as well as careful site selection and orientation (Streimikiene *et al.*, 2020).

Table 4.4 gives an assessment of energy efficiency measures in households in terms of cost efficiency and the potential for reducing GHG emissions.

As shown in Table 4.4, support measures for energy efficiency improvement have different potential cost-effectiveness linked to energy savings and the potential for reducing GHG emissions. It is important to underline that many of the examined measures produce the best result when combined. The highest economic efficiency and reduction of GHG emissions are characteristic to the following regulatory and control energy efficiency improvement measures in households: electrical equipment standards, especially when used with energy labelling and are kept under constant review, energy efficiency obligations and quotas, and demand management programmes;

TABLE 4.4

Assessment of energy efficiency policy measures in households

Policy measure	Reduction of GHG emissions	Cost-effective	Strengths, weaknesses and implementation requirements
Regulatory and control measures			
Standards for installations	High	High	Success factors: periodic updating of standards, independent control, information, communication, education
Energy certification of buildings	High	Average	Effective only when ongoing and periodically updated
Energy efficiency obligations and quotas	High	High	Continuous improvements are needed: new energy efficiency measures, short-term incentives to restructure markets
Mandatory audits	High but variable	Average	Most effective if combined with other measures such as financial incentives
Labelling and certification systems	Average / High	High	Mandatory schemes are more effective than voluntary schemes. Coordination with other measures and regular updates may increase efficiency
Demand management programmes	High	High	Generally, for the commercial sector, these programmes are more effective than in the housing sector.
Economic and market mechanisms			
Energy saving contracts (ESCO)	High	Average	Strength: State spending or market intervention is not needed, the benefits of increased competitiveness.
Collective procurement	High	Average / High	Harmonization with standards and labelling, choosing products with technical and market potential
Energy efficiency certificate schemes (white certificates)	Average	High / Average	There is no long-term experience. Transaction costs can be high. Institutional structures are needed. Full synergies with existing policies. Employment benefits
Kyoto flexible mechanisms	Low	Low	So far a limited number of CDM and JI projects in buildings
Fiscal and financial measures and initiatives			
Taxes: CO_2 fee or fuel taxes	Low	Low	The impact depends on price elasticity. Revenue may be allocated to further energy efficiency support schemes. More effective when combined with other measures
Tax incentives	High	High	If properly designed, encourage the introduction of highly efficient equipment into existing and new buildings
Public utility taxes	Average	High	Success factors: independent fund administration, regular monitoring feedback, simple and clear design

(continued)

TABLE 4.4 (Continued)

Assessment of energy efficiency policy measures in households

Policy measure	Reduction of GHG emissions	Cost-effective	Strengths, weaknesses and implementation requirements
Capital subsidies, grants and tax incentives	High	Low	Positive for low-income households, at risk of free movement may encourage innovative investment
Dissemination of information and voluntary measures			
Voluntary and negotiated agreements	Average / Height	Mean	Can be effective when implementation is difficult, combined with financial incentives and regulatory threats
Public leadership programmes, including public procurement regulations	Average High	High / Average	Can be used effectively to demonstrate new technologies and practices. Mandatory programmes have greater potential than voluntary
Education and information programmes	Low / Average	Average High	It is more used in the residential sector than in the commercial sector. Best to be applied in combination with other measures

Source: Created by the authors.

however, the latter have a higher level of climate change mitigation and cost-effectiveness in the commercial sector compared to the residential sector (Streimikiene *et al.*, 2020).

Among the most efficient and effective economic and market-based measures to increase energy efficiency are collaborative procurement programs; however, they are more effective when combined with equipment standards and labeling.

Tax incentives can be distinguished from fiscal initiatives as they are very effective in ensuring the deployment of high-efficiency facilities in new or renovated buildings.

Information and voluntary energy efficiency promotion measures, according to the experience of many countries, show lower efficiency in reducing GHG emissions and saving energy and public finances. Public leadership programmes, including public procurement regulations, are among the most effective measures; nevertheless, it must be stressed that mandatory public procurement programmes are significantly more effective than voluntary programmes.

Renovation of multi-apartment buildings is a very important measure of reducing GHG emissions, especially in the former socialist countries, which inherited a large stock of energy inefficient buildings. Promotion policy measures and schemes that can overcome various obstacles need to be carefully selected when making investment decisions for the renovation of apartment buildings (Banfi *et al.*, 2008; Sorell *et al.*, 2009).

The owners of multi-apartment buildings cannot make a joint decision on the refurbishment of multi-apartment buildings on a large scale due to the differences in their age, education, level of awareness, income, and also because they have different values and avoid risks and preferences. Also, there are risks related to time, environmental and social attitudes of apartment owners. Policy schemes should take into account the perceived risks of energy renewal and ensure that owners of apartments will generate future revenues as a result of energy savings. The ESCO model can be adapted to contracts for energy renewal in multi-apartment buildings in order to attract investments that are too risky for apartment owners. The UK government implemented the ESCO model more than 20 years ago (Crilly *et al.*, 2012); yet all the above-mentioned measures have their own weaknesses and are unable to fully overcome the barriers to energy efficiency improvement, for which they were implemented. Below we will discuss all energy efficiency improvement measures and assess their effectiveness in overcoming energy-saving barriers and qualitatively assess the GHG emission reduction potential and the cost-effectiveness of the measure. The main measures to promote the renovation of multi-apartment buildings can be divided into five broad categories: regulatory and control measures, information and voluntary measures, energy and CO_2 taxes, flexible market-based measures and financial incentives.

Table 4.5 lists policy schemes for the promotion of the renovation of multi-apartment buildings in different countries. Policy schemes are grouped into regulatory and information measures, market measures, fiscal and financial measures. An assessment of their effectiveness in overcoming the barriers to renovation has also been carried out.

An overview of the main policy measures to promote housing renovation and an assessment of the effectiveness of removing renovation barriers (Table 4.5) showed different results. In Table 4.5, the same grouping as in Table 4.4 was used for policy measures to promote the renovation of housing, analysing all measures to promote energy efficiency in households; however, regulatory and control measures were combined with educational and information measures, as the latter are closely linked and often applied together, making their distinction quite difficult for the promotion of housing renovation.

As shown in Table 4.4, regulatory and control measures and communication and information measures have been put in place in many EU Member States. Some regulatory measures are voluntary and some are mandatory. However, taking into account these policy schemes, behavioural and organizational barriers that prevent multi-apartment owners from making decisions of dwelling renovation have not been removed. These policy schemes require periodic updating and independent monitoring. In addition, education, information and feedback are necessary. Mandatory schemes are more effective than voluntary schemes. To improve the effectiveness of these measures, they must be combined with other incentives, fiscal or economic measures that simulate the market (Streimikiene *et al.*, 2020).

TABLE 4.5

Assessment of the effectiveness of policy measures for the promotion of the renovation of multi-apartment buildings

Measures and countries that have implemented them	Effectiveness of measures to overcome obstacles to renovation	Recommendations to improve effectiveness
Mandatory or voluntary regulatory or gradual regulatory or control measures; education and information measures (Germany, UK, Denmark, Italy, France, Sweden, England and many new EU Member States)	These measures do not remove barriers to multi-apartment owners to make decisions on the renovation of their apartments on a large scale.	Periodic renovation and independent control are essential. In addition, it is necessary to provide information, communicate and learn. Mandatory schemes are more effective than voluntary schemes. Efficiency needs to be combined with other strategies.
Market-based measures such as white certificates or energy performance certificates (Italy, France, Poland, England and Wales)	These measures do not eliminate organizational barriers when deciding on energy renewal in multi-apartment buildings. They have high monitoring and enforcement costs, they need institutional support and their interaction with other measures is unclear.	It is necessary to provide information, communicate and learn. In order to increase efficiency, harmonization of labelling and standards is essential.
Fiscal measures: CO_2 and energy taxes (Germany, Sweden, Norway, Denmark)	High energy taxes have an impact on household energy saving habits, but external energy production costs are not fully included and the impact is insignificant.	Efficiency is linked to price elasticity. Most effective when combined with other measures.
Financial measures: programmes providing subsidies and preferential loans for energy renewal (Germany, UK, France, new EU Member States)	They can provide access to the capital of low-income residents; nevertheless, the effectiveness of these programmes is poor as they are not well-targeted and for the majority of owners receiving a subsidy, who would have renovated their apartments without a subsidy.	Independent administration of funds, regular monitoring and feedback are necessary. Simple and clear design is necessary.
Policy packages designed to remove a variety of financial and other barriers at the same time (EU, US, Japan)	No results on the effectiveness of these policy packages in order to overcome the main barriers arising from large-scale energy renewal programmes.	

Source: Created by the authors.

Market-based measures such as "White Certificates" or "Energy Performance Certificates" have been put in place in several EU Member States; yet according to the experience of these countries, they do not remove behavioural and organizational barriers to decisions on dwelling renovation. These measures need to be combined with education and information measures and campaigns. In order to increase the effectiveness of these policies, it is also advisable to combine these measures with energy labelling and plant energy standards.

Fiscal measures, such as CO_2 and energy taxes, have been introduced in many EU Member States; yet only in Sweden, Denmark and Norway very high energy and carbon taxes influence changes in household behaviour. In other EU Member States, fees are too low to encourage residential renovation projects. These measures do not overcome important behavioural and organizational barriers to housing renovation projects and their effectiveness is linked to energy price elasticity. Generally, these measures are also more effective when combined with other measures such as regulation and control or information measures.

Financial measures, such as programmes providing subsidies and preferential loans for housing energy renewal, are the most popular policy schemes to support the renovation of multi-apartment buildings. Usually, residents face some economic barriers to house renovation, and in general, the effectiveness of these programmes is poor for a variety of reasons. For example, many owners of multi-apartment buildings receive this subsidy, not considering the fact that some would have renovated their homes without a subsidy. Moreover, these policy schemes are unable to address many organizational and behavioural barriers related to investment in housing renovation.

As the information provided in Table 4.5 shows, in order to successfully implement policy schemes to promote the renovation of multi-apartment buildings and realize the whole potential for energy savings, a number of important issues should be taken into account in the implementation of these policy schemes in order to ensure their efficiency by overcoming the main barriers to housing renovation.

Policy packages have been implemented in several EU Member States, the US and Japan, aimed at simultaneously removing various barriers to housing renovation. As already mentioned, many of the measures in order to be successful should be implemented in conjunction with other measures. This is particularly important for awareness-raising and voluntary action. Information measures will also yield more effective results if they are properly structured and implemented in conjunction with other measures. However, the literature analysis has shown that policy packages linking several policy schemes to promote the renovation of multi-apartment buildings have not yet produced very good results in reducing barriers to improving energy efficiency in multi-apartment buildings. All schemes should be implemented by assessing their interrelationships in order to achieve synergies in overcoming the specific barriers to the renovation of multi-apartment buildings.

4.4 Conclusions

The penetration of RES in the household market is hampered by social, economic, technological and regulatory barriers. All these barriers require targeted policies and specific measures aiming at individual sectors with a view to increasing the use of RES: energy, heat and cooling and transport. From an administrative or competitive point of view, key policy measures removing barriers to market penetration of RES are economic and financial measures (subsidies and grants, taxes, tax incentives, fixed electricity purchase prices, market instruments), regulatory measures (objectives and strategic planning, standards, obligations and powers, building standards and energy performance class labelling), information and education measures (permanent monitoring and invoicing, demonstration projects, information dissemination campaigns) and voluntary measures such as corporate social responsibility and voluntary agreements.

It can be observed that most European countries' policies promoting RES are based on setting targets: the determination of the share of RES in final energy consumption of energy, electricity, heat and transport, and the application of regulatory and control policies and fiscal and financial measures. Similarly, RES has implemented non-small education and dissemination measures to promote public interest and positive preferences for renewable energy technologies.

Although a high level of public understanding and acceptance of these technologies is required for the achievement of the goals for the use of RES, targeted policies and effective measures to remove social, behavioural and psychological barriers to the market penetration of RES are still lacking. EU countries need to take account of barriers to the use of RES in the transport sector and develop additional policies to achieve renewable energy targets in the transport sector. The example of Latvia, with the introduction of a vehicle tax, can also benefit other EU and Baltic countries facing problems in implementing the targets for the use of renewable energy resources in transport.

The process of energy efficiency improvement in households also faces important regulatory, economic, social and technological barriers, as well as the development of the use of RES. Regulatory and control measures, economic or market, fiscal and financial measures and educational and information measures are used to overcome these barriers.

Support measures for energy efficiency improvement have different potential cost-effectiveness linked to energy savings and the potential for reducing GHG emissions. It is important to stress that the implementation of these various packages of measures reaches the best results. The following regulatory and control measures have the highest economic efficiency and reduce GHG emissions: energy efficiency improvement measures in households: electrical equipment standards, especially when used with energy labelling and are kept under constant monitoring, energy efficiency obligations and quotas and demand management programmes; nevertheless, the latter have a higher level of climate change mitigation and cost-effectiveness in the commercial sector compared to the residential sector.

Collective purchasing programmes and tax incentives are among the most effective and efficient economic and market mechanisms and fiscal policy measures to improve energy efficiency. Energy efficiency promotion information and voluntary measures, according to the experience of many countries, show lower efficiency in reducing GHG emissions and saving energy and public finances. Public leadership programmes, including public procurement regulations, are among the most effective measures if they are mandatory.

The renovation of multi-apartment buildings has the greatest potential for energy savings and reduction of GHG emissions in households, but due to various barriers, such as economic/financial, informational, behavioural and organizational, it is not possible to realize the full potential of energy savings.

As demonstrated by an analysis of the experience of different countries, the current measures of the policies in place to overcome barriers to the renovation of multi-apartment buildings are the following: regulatory and control measures, information and voluntary activity, energy and CO_2 taxes; flexible market-based measures and financial incentives are not properly coordinated and oriented and therefore cannot overcome the main barriers to the energy renewal of residential buildings and therefore need to be combined and applied in a highly thought-out manner, particularly in order to overcome organizational and behavioural barriers to the renovation of multi-apartment buildings. Therefore, in order to develop effective policies for the promotion of housing renovation, further research is needed.

5

Policies to Support Renovation of Multi-Flat Buildings and Address the Energy Poverty of Ageing Societies

5.1 Improving Energy Efficiency in Buildings

Improving energy efficiency in buildings is a main way for reducing greenhouse gas emissions in Lithuania, as buildings account for more than 40% of the EU's final energy consumption. Though energy savings in residential buildings have huge potential, especially in Eastern Europe, where there is huge old and poorly built apartment building stock, the energy renovation process is very slow in these countries, including Lithuania. Energy renovation rates in Europe are estimated to be at around 1% per year (BPIE, 2012).

Variety of policies and measures were implemented in the European Union (EU) Member States; some measures are innovative; however, they cannot solve the problems that prevent the rapid energy renovation of apartment buildings having the greatest potential for energy savings and GHG emission reduction.

Thus, when developing policies and measures to encourage the energy renovation of apartment buildings, it is important to understand the perspectives of both building owners and institutional investors in order to put in place the most appropriate schemes to accelerate the energy renovation market for multi-flat buildings (Matschoss *et al.*, 2013). Most of energy renovation investments in residential buildings will be invested by the owners of apartments in residential buildings. Instead, governments can support and encourage such decisions through grants, but these public support funds can only cover just insignificant portion of the necessary investments for energy renovation. Therefore, additional measures may be important and necessary.

The main problems in the implementation of energy efficiency improvements in apartment buildings are related to the limited rationality of apartment owners, organizational problems, high transaction costs between owners of apartments and construction companies or project developers, as well as to dysfunctional markets in renovation supply chains. Besides that, there is a lack of research on the barriers of energy renovation, especially in post-Soviet countries.

Existing research on energy efficiency in residential buildings has emphasized on several technical, organizational, economic, financial and behavioural-psychological barriers to renovation of apartment buildings (Golove & Eto, 1996; Wittman *et al.*, 2006; Uihlein & Eder, 2009; Lujanen, 2010; Jensen & Maslesa, 2015). These barriers are particularly difficult to overcome for owners of apartment buildings in shrinking and ageing communities of post-Soviet states (Weinsziehr *et al.*, 2017). There are communities in new EU Member States, where the birth rate is falling sharply, the emigration of young households is growing and a large stock of low-quality, inefficient apartment buildings exist. Hence, in towns and neighbourhoods with a large elderly population, the problem of "empty nests" is often a main barrier to energy renovations of apartment buildings (Herfert & Lentz, 2007).

Several studies show that the intensity of renovation of apartment owners over the age of 70 is much lower than that of the younger population (Weinsziehr *et al.*, 2017; Zoric *et al.*, 2012). In Slovenia, Zoric *et al.* (2012) discovered that the probability of modernization of apartment buildings is negatively affected by the age of apartment owners. Nuisances and troubles caused by renovation activities are also important reasons for the reluctance of apartment owners to carry out energy renovation (Karvonen, 2013). Retirement homeowners face an additional hurdle because apartments are no longer handed over to family members because grown children have left the country and do not have plans to return (Weinsziehr *et al.*, 2017; Matschoss *et al.*, 2013; Ma *et al.*, 2012; Burger, 2013; Kilip *et al.*, 2014; Labanca *et al.*, 2015; Guerra-Santin *et al.*, 2017). By contributing to fuel poverty, the areas with the least energy efficient building stock and the highest heat consumption often overlap with elderly, retired, and low-income households (Herrero & Urge-Vorsatz, 2012; Boardman, 2010).

Lastly, an under-researched problem is that the owners of old apartment buildings were unable to agree on energy renovation of their buildings. While decision-making is a fairly simple process for owners of individual houses, it is significantly more complex when owners inadvertently depend on each other or have diverse and sometimes conflicting interests (Matschoss *et al.*, 2013). This is a significant problem for policies and measures aiming to modernize and improve energy efficiency in residential buildings as current schemes can't solve such type of problems.

5.2 Problems of Renovation of Multi-Flat Buildings in East Europe

In developing policies and measures to encourage the renovation of apartment buildings, it is essential to understand why many households are not currently involved in this process. Though there has been no comprehensive research in the EU that was directly asking households for their views on the main barriers to energy renovating their homes, there are several studies addressing the main barriers to energy renovation of multi-flat buildings.

The term "barriers to the renovation of apartment buildings" is related to the concept of the energy efficiency gap (Golove & Eto 1996; Jaffe & Stevins 1994), which means that investment in energy efficiency is significantly lower than desirable given the social, economic, environmental and technological optimum. This concept has been widely used to identify the main barriers to energy renovation faced by households, societies and countries.

The study by Itard *et al.* (2008) found that the main barriers to energy renovation of multi-flat buildings are a lack of knowledge, lack of information, a lack of funding and lack of cost-effective modernization schemes. Empirical studies of households living in older, low-quality, energy inefficient buildings have discovered many reasons why no renovation work has been undertaken, and have highlighted a variety of interlinked barriers to energy renovation: household values and attitudes; high price; poor organization; inconveniences; low-skilled, untrustworthy or overpriced professionals; lack of clear information on actual energy and savings; difficulties in making a decision for all apartment homeowners (Mallaband *et al.*, 2012). A wide-ranging study in five European countries identified similar barriers to energy renovation of multi-flat buildings: lack of information on projected energy savings, large expenses, weak political incentives and frequently changing public policies and incentives (Beillan *et al.*, 2011). EST (2010) conducted a survey of unrenovated apartment buildings and identified the following barriers to energy renovation: lack of appropriate targeted information and household awareness, inadequate motivation, major inconveniences, and low affordability. Similar findings were achieved in other works (Consumer Focus, 2012; Huber *et al.*, 2011).

There are also studies dealing with the drivers of energy renovation. The study by Watson (2011) provided an interview with homeowners about the motives of renovation. The study found that motivation is related to the context in which residents find themselves (Fawcett *et al.*, 2013). In Germany, which has strict policies of supporting energy renovation through low-interest loans and generous grant financing, the motivation for energy renovation is different from that in post-Soviet states like Lithuania (Novikova *et al.*, 2011). In general, the most important motivation in all countries is achieved thermal comfort and cost savings due to reduced energy consumption (Consumer Focus, 2012; Huber *et al.*, 2011; Nair *et al.*, 2010; Stiess *et al.*, 2009; Cadima, 2009).

Various studies have identified several groups of barriers to residential energy renovation. Jensen and Maslesa (2015) allocated barriers of energy renovation into internal barriers, mainly related to inertia of apartment owners, and external barriers related to lack of knowledge, resources and indentures among apartment owners in the same building.

The Better Buildings Partnership (2010) has sorted barriers of energy renovation into several main groups:

- Commercial barriers or market failure to deliver investment in energy
  renovation and the split incentives between landlords and tenants;
- Barriers of energy renovation processes, as there is a lack of defined
  procedures on how to define individuals having responsibility and
  authority to plan, organize and implement energy renovation of multi-
  apartment buildings;
- Financial barriers due to limited access to capital of apartment owners
  or third party;
- Technology barriers due to lack of knowledge about the best renovation
  options;
- Policy barriers due to ineffective regulation.

Uihlein and Eder (2009) presented the following barriers to multi-flat
buildings: uncertainties linked to costs and energy savings, financial barriers,
lack of information, lack of renovation skills, high transaction costs, various
organizational problems and several context-dependent barriers.

Uncertainties related to costs and savings are built on the basis that although
cost-effective solutions can be derived it is not necessary that the same benefit
can be achieved for all types of similar energy renovation as other issues can
also play an important role (Golove & Eto, 1996; Uihlein & Eder, 2009; de
T'Serclaes, 2007; de T'Serclaes & Jollands, 2007; Matschoss *et al.*, 2013). This
is due to the availability of conflicting information on the costs and benefits
of renovation, providing for the mistrust of such information (Golove & Eto,
1996). Indeed, the estimated average energy savings may not be achieved
by individual apartment owners due to specific building characteristics or
energy usage patterns. Energy savings achieved during renovation usually
depend on energy prices and interest rates so that the uncertainties linked
to verification of energy savings become barriers to renovation of multi-flat
buildings (Uihlein & Eder, 2009).

Financial barriers play an important role as significant capital investment is
necessary for energy renovation of residential buildings (de T'Serclaes, 2007;
Uihlein & Eder, 2009; Herrero & Urge-Vorsatz, 2012). Several studies have
highlighted the upfront costs as the major factor in the renovation of residential buildings. Other serious economic barriers include the lack of monetary
savings, lack of financial resources (Novikova *et al.*, 2011) and reluctance
of low-income households and older residents to apply for loans (Stiess
et al., 2010).

Due to the landlord/tenant dilemma, both tenants "ability to pay rent
and ability to suspend landlords" pre-investments are often directly
related to low household income levels (Weinsziehr *et al.*, 2017). Stiess

et al. (2010) reported that apartment owners with an income of less than €1,500 per month are less likely to renovate their apartments compared to owners with an income higher than €1,500 per month. Zoric *et al.* (2012) discovered that apartment owners older than 70 years show less willingness than younger apartment owners to renovate their apartments. The study by Zoric *et al.* (2012) provided that the probability of energy renovation of apartment buildings is negatively affected by the age of apartment owners. The disruptions due to renovation activities are other important barriers explaining the reluctance of households to carry out energy renovation activities in their homes (Karvonen, 2013). As mentioned above, retired apartment owners may face an extra barrier if their children have left home and do not intend to inherit the flat.

In addition, households use a simple "rule of thumb" for decision-making in energy renovation and choose measures with short-term payback period (Wittman *et al.*, 2006).

Besides that, some households do not have access to capital or there are high costs of borrowing for low-income households and they are reluctant to take a loan. Uncertain property prices are also significant barriers to energy renovation in the residential sector, as apartment owners anticipating to selling their property are not encouraged to renovate their apartments.

Lack of information and appropriate skills are other barriers to energy renovation of apartment buildings. Households not monitoring energy consumption in their apartments are reluctant to receive information about the possibilities of energy renovation and associated energy and cost savings (Uihlein & Eder, 2009; Golove & Eto, 1996). Besides that, there is a shortage of professionals and experts to provide this information to households. There is also the cost of obtaining and using information. Searching for relevant information or employing external experts to help in decision-making among energy renovation alternatives is also expensive, especially because the outcomes of energy renovation are quite uncertain for apartment owners. Transaction costs or so-called "hidden costs" are linked to the expenses of finding information and the expenditures associated with the monitoring of the contracted energy renovation work.

There are several logistical barriers like the lack of skilled providers of energy renovation services for households (Uihlein & Eder, 2009). Besides that, there are switching costs related to any changes in daily routine.

Organizational barriers are related to decision-making on the renewal of joint ownership, known as the principal-agent dilemma (de T'Serclaes & Jollands, 2007). With regard to the principal-agent dilemma, which is widely described in the scientific literature, lack of knowledge, information and funding are major barriers to private investment in energy renovation, as tenants will not benefit from investment in rented buildings and landlords will not benefit from it as they do not live in these warmer renovated apartments and tenants pay the energy costs of their rented apartments (Mallaband *et al.*,

2012; Beillan *et al.*, 2011; Dabia, 2010; Novikova *et al.*, 2011; Owens & Driffill, 2008; IEA, 2008 Witman *et al.*, 2006; Sorrel *et al.*, 2004).

The problem of collective solutions related to multi-apartment buildings is the most difficult to solve (Lujanen, 2010). This is more relevant in cases where the building is inhabited by different types of residents: part-owners and part tenants. Elderly people or tenants who expect to relocate soon are reluctant to undertake any energy renovations. Energy renovation of apartment causes major disruptions and can cause great stress, especially for the elderly who are accustomed to their daily routine. In apartment buildings with different households, organizational problems can take on even greater significance. Short decision-making time is also an important organizational barrier.

It is therefore necessary to emphasize the most important problem and obstacle of apartment house energy renovation. Collective decision-making is difficult to make due to housing renovation problems (Matschoss *et al.*, 2013). This is a problem that has so far received very little attention in the relevant scientific literature, despite being mentioned in several studies as one of the most critical barriers to energy renovation in the residential sector (Vainio, 2011; Matschoss *et al.*, 2013; Weinsziehr *et al.*, 2017).

Generally, decision-making is quite simple if there is only one owner of the building, but it becomes very difficult if there are several owners who are dependent on each other and who are forced to agree and make a joint decision on energy renovation of their common apartment building. Barriers to investment in energy renovation of multi-apartment buildings are particularly relevant in post-Soviet states, due to the large number of inefficient residential buildings. In EU new Member States this is also a big problem. Majority of the residents of multi-flat buildings have to make collective decisions but sometimes they have quite a lot of conflicting interests. Many residents of apartment buildings are not informed about the possibilities of energy renovation or their costs and benefits because they do not participate in residents' meetings. Some landlords fear that the renovating cost will not be reimbursed when selling their apartments (Nikola, 2011). Therefore, decisions on energy renovation can take several years and require numerous meetings and discussions among residents of apartment buildings (Matschoss *et al.*, 2013). The age structure, family size and income of apartment homeowners are the most problematic issues that are necessary to be addressed (Burger, 2013). Residents of the same building often earn different incomes, as in most apartment buildings in post-Soviet countries the size of apartments is very different – from one room to five rooms in the same building. Older and single people are reluctant to renovate due to low-income and "empty nest" situation (Vainio *et al.*, 2011). According to Weinsziehr *et al.* (2017), all households with adults over 45 years of age and with no children are classified as potential empty nests.

Energy poverty is related to ageing societies and the vulnerability of the lives of single retirees in East Europe countries, especially those countries suffering from high emigration, low birth rates and shrinking and ageing societies. Areas with the lowest building quality and insulation in these countries use large amount of heat, and low-income and older households are the main occupants of these buildings (Boardman, 2010; Weinsziehr *et al.*, 2017).

Therefore, the energy poverty problem is relevant in these countries due to high heating costs and energy inefficient buildings. This term is being used to describe households that cannot afford socially and materially necessary energy services such as heat, hot water and electricity (Bouzarovski, 2014; Bouzarovski & Petrova, 2015; Moore, 2012; Weinsziehr *et al.*, 2017). Rising energy prices and low household incomes associated with an ageing population in post-Soviet countries are exacerbating the problem of energy poverty. Low-income households tend to live in less energy-efficient, poorly insulated buildings because they cannot afford better housing or energy renovation (Bouzarovski & Petrova, 2015). As a result, low-income households spend more of their income on energy in comparison to high-income households (Hernández & Bird, 2010). This has increased the level of energy poverty in these countries. Even in Germany, the problem of energy poverty has enlarged in recent years as from 2000 to 2010 energy poverty rates increased from 12% to 18% in the country following the increase in energy prices (Weinsziehr *et al.*, 2017). Therefore, the issue of energy poverty is not attributed to the post-Soviet legacy (Herrero & Urge-Vorsatz, 2012). Given the differences in age, income and other characteristics of apartment owners in the same building, it can be observed that the decision to renovate is very difficult, especially if this decision requires the approval of a large majority of apartment owners. In Romania and Bulgaria, acceptance of 67% apartment owners is required (IIBW, 2008; Matschoss *et al.*, 2013), whereas in Germany and the Czech Republic, acceptance of 75% apartment owners is required (Burger, 2013). According to Heiskanen *et al.* (2012), most EU Member States, like Austria, Finland, France, Spain and Italy, generally require the consent of more than 50% of apartment owners.

If the owner of an apartment does not agree for renovation of a multi-flat building and does not share the cost of renovation, then other apartment owners must pay for the renovation themselves and only later on, in court, sue to be reimbursed. Currently, there are no other ways to overcome this barrier. When an owners' association decides to undergo a large-scale, high-cost renovation, the decision to renovate should be approved by all owners. In addition, some multi-flat buildings do not have an owners' association or other central body able to organize and carry out renovation of buildings.

Various categories of barriers to the renovation of apartment buildings overlap and reinforce each other. There are many risks and uncertainties associated with energy renewal, which may put apartment owners at even greater risk than they are in the present situation, given the current and underdeveloped market for energy renovation and related financial services. Most importantly, the main reasons for inactivity are multifaceted and interrelated, depending on the type of household, and vary all the time (Rittel & Webber, 1973; Ritchey, 2011).

Given the high complexity of this problem the policies and measures to encourage the renovation of apartment buildings in the light of the above barriers will be discussed and the adequacy of existing policies and measures to overcome these barriers will be assessed.

5.3   Assessment of Policies and Measures to Overcome Barriers of Energy Renovation

Energy renovation of apartment buildings is the main method to achieve significant energy efficiency improvements in households. Energy efficiency improvements in multi-flat buildings are based on reducing heat consumption for both space heating and domestic hot water supply. Therefore, energy consumption in residential buildings can be reduced by increasing the energy performance of buildings through advances in thermal envelope of buildings and improvement of heating systems. Usually, such improvements in insulation and heating provide significant energy savings in residential buildings and are the targets of large-scale retrofitting of multi-flat buildings.

A review of policies and measures to promote energy renovation of apartment buildings indicated that these measures have provided for limited success in such countries as Germany, the UK, Denmark, the US and Japan based on the conducted studies (de T'Serclaes, 2007; Grösche & Vance, 2009; Galvin, 2015; IEA, 2008; Christense *et al.*, 2011; Charlier & Risch, 2012; Crilly *et al.*, 2012; Organ *et al.*, 2013; Kiliip, 2013; Burger, 2013; Horne & Dalton, 2014; Friege & Chappin, 2014; Labanca *et al.*, 2015; Ozarisoy & Altan, 2017; Sirombo *et al.*, 2017).

The key policies and measures to encourage renovation of apartment buildings can be classified as follows: information and advice, energy and pollution taxes, economic and financial incentives, access to capital and provision of minimum standards.

Information plays a key role in supporting households and aims at overcoming information failures available in the market. In the presence of asymmetric or incomplete information or both there are high uncertainty and high hidden costs faced by households. There are many programmes to provide information on energy renovation and advice but none of them have been able to reinforce energy renovation of apartment buildings. Energy certificates and labels developed to deliver information to customers are not effective in signalling about the benefits of renovated buildings and are not capitalized into the prices of renovated apartments yet. For EU countries energy performance certificates (EPC) are mandatory and a new directive for buildings; however this is not the case for renovated buildings as studies dealing with energy labelling in EU provided that they are even unknown to residents of various EU countries (Christense *et al.*, 2011). This is because homeowners think that energy certificates and labels deliver just trivial information (Christense *et al.*, 2011).

Feedback programmes are also significant advice tools. Energy audits are valuable tools because they ensure the availability of personalized information on energy consumption in homes, though empirical results suggest that the effectiveness of energy audit schemes varies considerably. Some works have found that energy audits can lead to some energy savings, but others do not show any impact or even exhibit the increased energy consumption (Sirombo, 2017). Energy bills provide information on household energy consumption and can be used to provide advice and comparable information. Smart metering or online monitoring systems offer a good opportunity to ensure feedback with households as they can provide continuous and real-time information. The installation of a building monitoring system (Sirombo *et al.*, 2017) or metering devices that track "voluntary" electricity, heating, cooling and hot water consumption in individual apartments is a valuable tool to ensure feedback as well. Nevertheless, the impact of feedback programmes on the decision to renovate is restricted by other barriers to energy renovation of apartment buildings.

Energy or CO_2 taxes that aims to reduce energy consumption by sending price signals to households and stimulate them to implement energy efficiency improvements are introduced in many countries. Well-designed taxes can integrate the external costs of energy consumption (such as environmental impact) into consumer decision-making. Some EU countries like Germany or Scandinavian countries having very high energy efficiency standards and high environmental and energy taxes have populations that are very familiar with energy savings. However, in other, less developed states with low incomes, taxes do not encourage renovation. In addition, low-income households in post-Soviet countries are entitled to VAT rebates for district heating, which is further discouraging energy renovation.

There are financial incentives in the form of supplier commitments, tax breaks, grants and soft loans that aim to encourage apartment owners to implement measures that reduce energy costs; however, their impact on apartment owners' decision to renovate apartment buildings has not been proven empirically. Energy Company Obligations (ESCOs) require energy suppliers to improve the energy performance of residential buildings and have been widely applied in EU; nevertheless these obligations have not necessarily achieved their objectives. It is believed that supplier commitments or ESCOs can save energy at low costs because energy suppliers have a competitive incentive to deliver measures as cheaply as possible (Labanca *et al.*, 2015). They also have the advantage of working with suppliers who have marketing skills and have several well-established contacts with customers in apartment buildings. ESCOs together with the white certificates were applicable in the UK, Denmark, France and Italy. There was a plan to implement ESCOs in Poland; however, these schemes are not so popular in other post-Soviet countries due to the specific of Soviet district heating systems mentioned above.

Implemented ESCOs feature a wide variety of designs and provide for diverse outcomes in specific countries. The UK ESCO model focuses on the implementation of costly renovation measures such as wall insulation and is targeted at low-income households with specific commitments (carbon saving, carbon saving communities and affordable heat obligations). The UK ESCO was set up in 2013 to reduce energy consumption and combat fuel poverty. The scheme was scheduled to continue until 2017. The ESCO in the UK sets out legal obligations for energy suppliers to implement advanced energy efficiency measures in households and works in conjunction with the Green Deal. The aim of the Green Deal is to support low-income households in energy renovation of their flats by making it possible not to pay for renovation costs in advance, but to cover their monthly utility bills for energy. A Green Deal would be an effective way to tackle energy poverty and a low-income, ageing population; however, empirical data suggest that this scheme is not the best means of saving energy in residential buildings and is very expensive compared to the energy savings achieved (Christense *et al.*, 2011). Furthermore, the scheme does not remove obstacles to decision-making in the energy renovation of multi-apartment buildings. Consequently, the ESCO model needs to be adapted to their national circumstances.

In addition, the majority of energy savings to date have come from low-cost measures that improve energy efficiency, with the savings spread across a large number of households. Going forward it will become more challenging for obligations to provide energy savings as the potential for additional low-cost measures falls over time. Thus, a body of evidence from EU Member States suggests that supplier obligations are not the ideal tool for supporting renovation of multi-flat buildings, as they can be regressive since the costs are recovered through energy bills (which are a greater proportion of poor households' income) and suppliers can deliver their obligations at least cost by providing measures to higher-income households that can contribute towards them (Janda and Parag, 2013).

Grant schemes that provide measures at reduced or zero cost to households clearly encourage uptake but are costly to taxpayers. They have, therefore, tended to be targeted to particular users, e.g. towards fuel-poor households. However, a problem is that grant schemes that focus on delivering improvements to fuel-poor households can have a significant rebound effect, where households respond to the improved efficiency by heating their homes more, which can limit energy savings. In addition, grant schemes often suffer from free ridership, where subsidies are provided to households that would have completed the upgrade even without the subsidy (Janda & Parag, 2013).

Tax discounts usually compensate households for part of the cost of energy renovation. They work as grants; however, tax breaks can be less effective, as there is the problem of free riders. Because tax discounts are usually reimbursed to households upon termination of renovation work, they are often found to be taken by higher-income households that can easily cover upfront costs of energy renovation. Similarly, a reduced VAT to 5% in the UK and to 9% in Lithuania applied to residential district heating in the end is providing greater support to higher-income sections of the population who have large apartments and use more heat.

Policies and measures to help to access the capital like loan guarantees and soft mortgage rates can ease energy renovations that are very costly in advance and may be particularly suitable for low-income homeowners with limited access to capital. Nevertheless, the lack of funding is not the main reason that hinders energy renovations of apartment buildings.

Loan schemes are usually "relaxed" by offering zero or low interest rates. They are usually realized through public-private partnerships, with the government guarantees for the bank and requirements set for banks to offer a reduced interest rate on loans granted to residents for energy renovation. An alternative way is to provide a guarantee by the government to share credit risk with banks to increase private investment uptake. Loan guarantees are applied where banks have sufficient liquidity but a relatively low risk margin (Healy, 2004; Janda & Parag, 2013; Janda and Parag, 2013).

Building codes are widely applied to regulate the standards to which buildings must comply. Building codes and electric appliances standards are very useful as they aim to take the least efficient products off the market, reframing consumer choices. There is strong evidence that standards deliver energy savings, although usually well below engineering estimates for buildings (Healy, 2004). Regulations setting out the minimum performance standards for buildings are common internationally for new-build properties, with standards raised over time to drive performance. Managing the performance gap between modelled performance and actual performance is important in making such standards effective. Enforcement is also needed to ensure regulations are adhered to. These factors can be more difficult at building scales where the impact of a technology can vary based on its situation. Many countries also use standards for energy renovation like Denmark, Sweden and Germany; however, there is still no empirical data about the effectiveness of these policies (Janda & Parag, 2013).

All the measures examined do not produce effective results for energy renovation of multi-apartment buildings, as these schemes do not adequately remove the main obstacles analysed above. If, from a household perspective, programmes are transparent and simple enough, then they are more likely to achieve the intended impact. However, there is one appropriate energy renovation policy for all cases and all countries due to overly complex barriers and country-specific situations. A special scheme should be developed in order to take into account the needs of different actors in the energy renewal process. It is also important to make steady progress in energy renovation to boost consumer confidence. It is essential that reliable intermediaries coordinate energy renovation programmes and manage their implementation effectively by dealing with organizational issues as well. Many countries have a designated specific energy agency to perform this function. There is also a need for a qualified supply chain for energy renovation services. For example, Germany has developed an energy renewal supply chain for reliable and highly qualified energy expert engineers (Janda *et al.*, 2014).

The way in which these policies are presented can increase consumer confidence. Targeting policies in specific geographic areas can improve engagement in broad social networks that spread message and recommendations on energy renovation to trusted parties (family members, neighbours, friends, etc.). The establishment of organizations that consumers already communicated with can help to build trust (Janda *et al.*, 2014).

Communication and marketing skills are important for policies to work. Communication will yield the best results when it conveys a direct and simple message to the target group, applying a simplified approach to all policies that support energy renewal (Organ *et al.*, 2013). Besides that, the success of a policy is bigger if it is coordinated with other policies designed to address similar or linked issues, but sometimes overlaps can lead to inefficient allocation or waste of resources (Charlier & Risch, 2012).

Other strategies to promote the introduction of renewable micro-generation technologies such as feed in tariffs (FITs) or feed in premiums (FIPs) are also related to the energy renovation of apartment buildings and need to be considered together (Kiliip, 2013).

It must be emphasized that the policies and measures described above cannot solve the problems of energy poverty and the difficulties of different income owners living in multi-apartment buildings in making a collective decision on energy renovation. Additional measures are necessary to encourage low-income households living in energy poverty. Supporting the renovation payment by increasing monthly energy bills of the low-income segment can prevent households from fuel poverty. Also, the achieved savings from energy renovation can be spent by households on other goods instead of investing in energy efficiency (Healy, 2004; Boardman, 2010). The other discussed measures to help low-income households and fight energy poverty by reducing the high cost of district heating such as VAT on district heating applied in Lithuania are not recommended due to their limited coverage, high administrative costs and the problem that higher income households with large apartments and consequently higher energy consumption will benefit more from such support. Such types of measures are called environmentally harmful subsidies and have to be abolished as they just distort markets and divert financial resources away from long-term solutions such as energy renovation of apartment buildings (Boardman, 2010; Healy, 2004).

For addressing energy poverty, in particular the related problems of collective decision-making, a new conceptual framework is necessary by taking into account the differences between apartment owners in terms of household size, age, habits, income, education and knowledge. The targeted policies and measures should be developed in the light of the shortcomings of current measures to reduce energy poverty, ageing and shrinking population in East Europe. In the next section an innovative framework to support energy renovation and overcome energy poverty issues is developed.

5.4 Innovative Scheme to Promote Energy Renovation of Multi-Flat Buildings

Energy affordability is a key aspect of energy services to households and various schemes can be applied to ensure the supply of energy to low-income households. It is often conflict between efficiency and equity than setting pricing schemes for energy and other services like water and heat, communications, etc. The efficiency principle in market economy requires that utilities prices should be set on the basis of long-run marginal supply costs and that all customers having the same long-run marginal costs (LRMCs) should pay the same price. The equity principle requires that prices should be linked also to affordability; thus the lower income households should pay less per unit of service than the higher income households. This conflict is not a new thing in decision-making. Even more developed countries provide some assistance to poor households to cover utilities costs. It is common in developing countries to have block tariffs or so-called life-line rates where for the lowest consumption levels the lowest price is set and each block faces the higher price for utilities consumption. Life-line rates are popular in developing countries and many of them have established such tariffs for utilities with one or more blocks.

Taking into account the principle of life-line rates the welfare costs of such scheme are shown in Figure 5.1.

There are two general rates for the two types of customers according to this scheme in Figure 5.1. The two types of consumers are as follows: "poor" and "rich". When the energy price is equal to LRMCs, poor households consume q_{po} and rich households consume q_{ro}. quantities of the good. The solution is effective in the sense that it increases consumer surpluses and no household can be put in a better position when other households are not doing worse. This solution is "Pareto efficient", according to the Welfare Economics terminology.

This Pareto efficient solution may be unsatisfactory for two reasons. First the share of income spent on energy by poor households may be too high, leaving them without enough money for other necessities. Second, the amount of energy they consume (q_{po}) may be considered insufficient for their social needs (Markandya & Streimikiene, 2003).

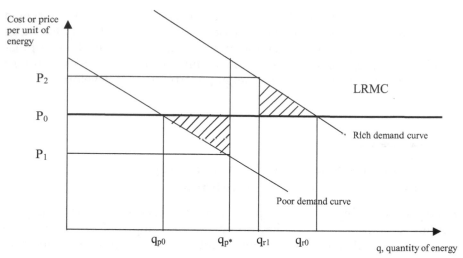

FIGURE 5.1
Welfare pricing of energy. Created by author based on Markandya and Streimikiene (2003).

The analysis of life-line rates principle in energy supply can be carried out in the following way. Suppose that the socially desired level of energy consumption for poor customers is q_{p*}. At the same time the energy utility must cover the costs of supply. This may be achieved by

(a) having a life-line rate of P_1 for energy consumption up to q_{p*};
(b) having a life-line rate at level P_2 for energy consumption above q_{p*}.

The welfare losses due to these changes are provided by the shaded areas in Figure 5.1, viz.:

$$\frac{1}{2}\times\left(P_0 - P_1\right)\times\left(q_{p*} - q_{P0}\right) + \frac{1}{2}\times\left(q_{r0} - q_{r1}\right)\times\left(P_2 - P_0\right) \tag{1}$$

However, the tariffs have to be designed so that the revenues of energy utility are equal to the costs of energy supply:

$$2P_1 \times q_{p*} + \left(q_{r1} - q_{p*}\right)\times P_2 = P_0 \times\left(q_{p*} - q_{p1}\right) \tag{2}$$

The design of the life-line rates is to ensure that the overall energy expenditure for poor customers is below a certain level:

$$\frac{q_{p*} \times P_0}{Y_p} \le \alpha \tag{3}$$

where Y_p is the income of poor customers and α is the declared acceptable share of energy expenditure.

The design problem may therefore be stated as follows:

Select P_1, P_2 tariffs for energy supply such as to minimize (1) subject to (2) and (3).

Besides doing the analysis with quantitative data, it is not possible to say completely what the exact welfare costs of this life-line scheme will be. Though there is some indication that these costs will be moderate due to the fact that consumer surplus gains and losses to different customers groups are not of equal value in welfare terms. According to the life-line scheme, poor customers will gain consumer surplus and rich customers will lose consumer surplus. The energy utility is left neutral. According to Welfare Economics, a dollar of loss to the rich is valued less than a dollar of gain to the poor; therefore the welfare costs of the scheme should be quite small (Markandya & Streimikiene, 2003).

Therefore, based on the assumptions above a similar life-line rates approach can be applied to energy modernization schemes that are too costly for low-income households, especially taking into account energy poverty issues in East Europe described above.

However, it is necessary to investigate household's willingness to share costs of energy renovation and to participate in such schemes, as energy renovation of multi-flat buildings is a complex process and there are many barriers linked to organizational problems and high capital costs. The proposed life-line rates approach combined with the ESCO model can provide good results in dealing with these main barriers to energy renovation in East European countries.

The ESCO model can solve organizational issues (Paiho *et al.*, 2015). The main revenue of the ESCO business model is related to the achieved reduction in energy costs or energy consumption. The ESCO model can provide all energy renovation services for multi-flat buildings: construction, engineering, financing, ensuring all permits and collecting contracts from apartment owners. The ESCO model may also include energy audit, consulting, operation and maintenance services after renovation.

As raising funds for energy renovation of apartment buildings is an very important issue, the ESCO model can apply invoicing of apartment owners and charging for energy renovation in monthly utility bills (Bell *et al.*, 2011). Energy renovation will make it possible to reduce the overall utility bill for energy savings (Würtenberger *et al.*, 2012). It is necessary to stress that energy renovation programmes can be successful when they are straightforward and contractors are able to quickly collect payment for their services (Johnson *et al.*, 2012; Paiho *et al.*, 2015).

Therefore, utilities in the district heating sector can apply the ESCO model and invoicing of apartment owners on energy bills by applying life-line rates for their services. This would allow to ensure the financial viability of large-scale energy renovations and address energy poverty and organizational barriers of energy renovation (Boute, 2012).

The case study conducted in Lithuania on willingness to pay (WTP) for energy renovation also evaluated the willingness to share energy renovation costs with low-income households in the same multi-flat building and evaluated preferences of households in terms of payment mode for energy renovation. The study indicated that households are willing to pay for energy renovation by invoicing them in monthly energy bills; however, the willingness to share costs of energy renovation with poorer neighbours in the same apartment buildings was very low. Nevertheless, more studies are necessary to investigate these issues deeply.

5.5 Conclusions

There are many barriers preventing energy renovation of multi-apartment buildings in East Europe: lack of access to capital, lack of knowledge about energy renovation, split initiatives, difficulties in reaching a collective decision on energy renovation among owners of apartments. These barriers are increasingly becoming more difficult to overcome in East Europe due to increasing fuel poverty of ageing, shrinking populations and huge stock of energy inefficient multi-flat buildings constructed during the Soviet era.

To overcome the barriers of energy renovation and to address energy poverty, new innovative policies are necessary. However, the most important issue is to solve the problem of collective decision-making, taking into account the different demographics of apartment owners residing in multi-flat buildings.

The ESCO model or on-bill financing model can be applied for the renovation of multi-flat buildings, based on the UK example. Higher payments for utility bills can be shared among households living in multi-flat buildings that require renovation. Life-line tariffs can be applied to pay for ESCO services. This enables sharing the costs of renovation among apartment owners having different incomes and addresses the principle of equity.

For implementation of this scheme, it is necessary to assess the willingness of apartment owners to pay for energy renovation of multi-flat buildings and to share costs based on income. The case study on WTP for energy renovation of multi-flat buildings performed in Lithuania provided very low willingness to share costs of energy renovation with poorer neighbours; however, future research is necessary for assessing the willingness of households to pay for energy renovation of multi-flat buildings in order to develop life-line tariffs for energy renovation in the ESCO on-bill financing model.

Conclusions

1. Although households consume about 30% of the final energy in Lithuania, the main climate change mitigation policy measures in the energy sector are directed to the energy supply sector. The approval and participation of the population are also necessary in formulating climate change mitigation policy. Therefore, climate change mitigation benefits are subject to households' willingness to pay (WTP), which reveals the societal priorities and its member's voluntary contributions to climate change mitigation. Thus, climate change mitigation policy in the energy sector should be based on an assessment of the population's WTP for specific climate change mitigation measures.

2. There are two main ways to mitigate climate change in households linked to energy consumption: the use of renewable energy sources or micro-generation technologies based on renewable energy sources, and increasing energy efficiency, with the greatest potential for energy savings available in energy renovation of multi-flat buildings. Therefore, when formulating climate change mitigation policy, the state must take into account the preferences of the population and their WTP for micro-generation technologies and energy renovation of housing. This is particularly important for selecting appropriate incentive schemes, as current financial initiatives such as subsidies and loan incentives, as shown by behavioural economists' studies and practices, are not the most appropriate way to promote appropriate behavioural changes mainly due to psychological barriers. Also because of wide range of market failures and various economic, social, regulatory barriers, the implementation of other climate change mitigation measures in households does not provide the expected GHG emission reductions.

3. After the analysis of studies on WTP for climate change mitigation in the energy sector, the scientific literature was allocated to two main groups: WTP for renewable energy sources and WTP for energy efficiency improvement. The scientific literature on WTP for climate change mitigation in the energy sector is dominated by studies on WTP for renewable energy sources. The literature on WTP for energy efficiency improvements in households is limited, although energy savings provide not only external benefits as in the case of renewables but also allow direct costs savings for households due to lower energy bills.

4. It was found that the choice experiment and the contingent valuation method (CVM) were the main attributes for assessing WTP for climate change mitigation measures in the studies examined. The studies on WTP for micro-generation technologies using renewable energy sources in households were mainly based on the choice experiment. The main attributes in the choice experiment for estimating WTP for micro-generation technologies were the cost of electricity or heat or the monthly bill for electricity or heat and the reduction of pollution (climate change mitigation). Other attributes used in the studies were the impact on local occupancy, sharing capability, device dimensions and the ease of equipment usage.

5. The main socio-demographic factors that determine the WTP for renewable micro-generation technologies were age, gender, education, income, price, position held, geographical location and environmental awareness. Only a few studies found that affiliation with environmental organizations, contract duration, political views and expected health impacts had an impact on WTP for renewable micro-generation technologies in households.

6. In assessing the WTP of the population for energy efficiency measures in households, the CVM and choice experiment were applied, by examining the WTP for certain measures or by establishing preferential support schemes. Support schemes and energy efficiency measures were described in attributes, without specifying them. The main attributes for assessing the WTP for energy efficiency improvement in choice experiments were annual energy savings, capital investment, reduction of CO_2 emissions, amount of state financial support and duration of soft loan.

7. The main socio-demographic factors influencing the population's WTP for energy efficiency measures in households were income, gender, age and education. Factors such as marital status, political views, membership of environmental organizations and positions held did not have a significant impact on household WTP for energy efficiency measures such as energy renovation.

8. A review of studies dealing with WTP for energy renovation revealed that households with housing renovation experience opted for more expensive energy renovation alternatives than households without such experience. In addition, it is natural that residents of lower energy-efficient housing have also opted for more expensive renovation alternatives, hoping to reap greater benefits from energy renovation.

9. Research has revealed that the size of the premium that residents are willing to pay for green or passive buildings is strongly and negatively affected by short-term financial initiatives taken by the government such as subsidies and loan incentives. These initiatives result in lower WTP for specific energy efficiency improvements in energy renovations.

This negative impact of financial support measures can be explained by behavioural economics insights that material incentives to promote desired behavioural change (energy savings) are ineffective and do not stimulate the desired behavioural changes.

10. As behavioural economists have shown, the media could make a significant contribution to shaping public opinion on energy efficiency measures and renewable energy sources through the disposition effect. Positive media views on renewable energy and the benefits of energy renovation (both financial and non-financial) will lead to higher WTP for these climate change mitigation measures in households and in the future, with growing environmental concerns and knowledge about the full benefits of renewable energy and energy renovation the need for support of these technologies and measures will decrease.

11. The CVM and choice experiments are the main methods of stated preferences that are used to evaluate the benefits of public goods using a hypothetical market. CVM and choice experiments differ in the form of questionnaires and different presentation of descriptions to respondents in assessing their WTP. In assessing the WTP by applying CVM, the change in the scale of a public good is described in general terms, without detailing any of its characteristics, and respondents are then asked to indicate their WTP a certain price premium for the indicated change. Meanwhile, the choice experiment allows the evaluation of several public goods, while the CVM is limited to only one alternative. The choice experiment is an indirect way of assessing WTP as it allows respondents to assess the WTP without naming a specific public good, but only providing its attributes (pollution reduction, employment growth, inequality reduction, etc.), which reduces the possibility of biased responses, especially in selecting between micro-generation technologies and support schemes. The choice experiment is more suitable for the assessment of WTP for renewable energy sources, and the CVM is more suitable for the assessment of WTP for specific energy efficiency improvement measures including energy renovation.

12. Two WTP for climate change mitigation studies were performed in Lithuania having the greatest potential of reducing greenhouse gas emissions: introduction of micro-generation technologies using renewable energy sources in private homes and energy renovation in multi-flat buildings. The choice experiment method was applied to WTP for renewable energy micro-generation technologies in households and direct CVM was applied to assess WTP for energy renovation of multi-flat buildings.

13. In the first survey, respondents are interviewed by providing them with various values of micro-generation technology evaluation criteria in a choice experiment-based questionnaire. Although the popularity

of micro-generation technologies among households is constantly growing, and most of the research confirms that people would like to pay extra for renewable energy, the situation in Lithuania is slightly different.

14. The study showed that Lithuanian households would pay extra only for such micro-generation technologies as solar collectors and solar thermal power plants, and the most important attribute determining the increase in their welfare is guarantees, but in all other cases households would like to receive compensation to choose micro-generation technologies such as a biofuel boiler or wind farm as WTP for these technologies is negative.

15. Given the estimates of the WTP, the four options can be ranked as follows (in terms of descending utility): solar panel, solar thermal installation, biomass boiler, and micro-wind. Again, the negative values of WTP should be used only for indication purposes rather than be strictly perceived as a need for additional financial support.

16. The differences in WTP for the four renewable micro-generation technologies have certain implications for the energy policy in Lithuania. Indeed, the country has various support measures for renewables besides the feed-in tariffs, i.e. financial support from EU Structural Funds and other funds, pollution tax exemptions for stationary pollution sources (for burning biomass), etc. The positive WTP for technologies based on solar energy (solar thermal installation and solar panels) indicates that these technologies are likely to prevail in the Lithuanian market. This is a suitable option for most of the urban dwellings; yet the situation might need certain corrections in the case of rural areas. Specifically, rural areas might exploit crop and forest residues by installing biomass boilers. Furthermore, micro-wind generators can also be an option for rural residents with abundant land areas. Therefore, the support for renewables can be differentiated across rural and urban areas by taking into account household preferences and WTP.

17. The possibility of sharing micro-generation technologies also proved to be insignificant for Lithuanian households. This can be explained by certain factors related to the Lithuanian context, in particular the negative experience of collectivization, which often inhibits cooperation initiatives between Lithuanian households, but deeper research is needed to substantiate this hypothesis.

18. The study of WTP for energy renovation using the CVM revealed that the absence of a housing association and the inability of neighbours to agree on the renovation of multi-flat building are the main organizational barriers of energy renovation in Lithuania. In addition, more than 90% of the respondents believe that the person or organization in charge of all organizational issues related to the energy renovation of

an apartment building would have a significant impact on their decision to renovate their homes.

19. More than 80% of the respondents believe that government support for energy renovations is insufficient and does not create the right incentives for energy renovation in multi-flat buildings, and a vast majority of households are dissatisfied with the level of thermal comfort and the claim that heating bills do not meet the level of thermal comfort. Respondents confirmed that they need renovation and would like to pay for the renovation by including the renovation costs in their monthly heat bills, rather than paying all the renovation costs immediately and paying the loans. In addition, more than 30% of the respondents are reluctant to take loan for energy renovation which is necessary taking into account their income.

20. The results of the survey showed that more than 90% of the respondents would not want to share the renovation costs of low-income neighbours and thus speed up the renovation process.

21. The penetration of renewable energy sources in households is hampered by the social, economic, technological and regulatory barriers. Overcoming all these barriers requires targeted policies and specific measures. The main policy measures needed to remove the barriers to the penetration of renewable energy sources are: economic and financial measures (subsidies and grants, taxes, tax incentives, administratively or competitively fixed purchase prices for renewable electricity tradable certificates, etc.), regulatory measures (strategic planning and targets for renewables; standards, obligations and mandates; building standards and energy labelling), information and education measures (continuous monitoring and itemized billing, demonstration projects; information campaigns) and voluntary measures, such as corporate social responsibility, and voluntary agreements.

22. It can be seen that most European countries' renewable energy policies are based on setting targets: the share of renewable energy sources in final energy, electricity, heat and transport, and regulatory and fiscal or financial measures. These countries have also implemented a number of education and dissemination measures to stimulate public interest and positive preferences for renewable energy technologies.

23. Although achieving the targets for the use of renewable energy sources requires a high level of public understanding and acceptance of these technologies, there are still no targeted policies and effective measures to remove the social, behavioural and psychological barriers to the penetration of renewable energy sources. EU countries need to address barriers for the use of renewable energy sources in the transport sector and to develop additional policies to meet renewable energy targets in the transport sector.

24. Improving energy efficiency in households also faces important regulatory, economic, social and technological barriers, as does the development of renewable energy sources. Regulatory and control, economic (fiscal and financial), and education and information measures are in place to overcome these barriers; however, these measures are not always effective.

25. It is important to emphasize that the best outcome is achieved through the implementation of these various packages of climate change mitigation measures. The following regulatory and control measures have the greatest cost-effectiveness and reduction of greenhouse gas emissions in households: standards for electrical equipment, especially when applied with mandatory energy labelling; energy efficiency commitments and quotas, demand management programs, but the latter are more effective in mitigating climate change in the commercial sector compared to the residential sector.

26. Collective purchasing schemes and tax incentives are among the most efficient and effective policies for energy efficiency improvement. Information dissemination and voluntary measures to promote energy efficiency, as the experience of many countries shows, are less effective in reducing greenhouse gas emissions and saving energy and money for the population. Public leadership programs, including public procurement regulations, are among the most effective ways if they are mandatory.

27. Renovation of apartment buildings has the greatest potential for energy savings and reduction of greenhouse gas emissions in households, but there are many barriers to energy renovation of apartment buildings, such as limited access to funds for apartment owners, lack of knowledge about large-scale energy renovations and inability of the apartment owners to agree on the renovation of multi-flat buildings. These obstacles are becoming increasingly difficult to overcome due to increasing energy poverty in an ageing society, especially the declining populations in East European cities.

28. As an analysis of the experience of different countries has shown, though policies are currently in place to overcome barriers to multi-apartment energy renovation such as regulatory and control measures, information and volunteering, energy and CO_2 taxes, and flexible market measures, these incentives are not well coordinated and targeted. All these measures need to be combined and applied very carefully, in particular to overcome organizational and behavioural barriers of multi-flat energy renovation, and further research is needed to develop effective policies to promote housing renovation.

29. In order to overcome the main barriers to renovation, it is necessary to apply both top-down and bottom-up strategies that promote the energy renovation of apartment buildings. The development of new

packages of policies and policy reforms to encourage the energy reno-vation of apartment buildings is an urgent priority for European Union in order to meet its 20-20-20 targets. Current top-down strat-egies initiated by governments can support and encourage invest-ment in the energy renovation of apartment buildings through tax credits and grants, but these available public funds can provide only a small part of the required investment and are therefore insufficient. Targeted, balanced and attractive innovative schemes are needed to give this market more impetus. Enabling a structured and regulated decision-making process and a single key player to organize all energy renovation work at multi-flat buildings could facilitate the renovation of multi-apartment buildings on a bottom-up basis.

30. Other innovative practices involve "middle actors" in the value chain of building renovation. Mid-level actors are important in the energy renovation supply chain, as they affect all other actors in the renova-tion process, including policy makers as well as energy producers and consumers. Therefore, energy renovation professionals in the building sector can influence the implementation of new policies and schemes and to ensure secure financing and attract new players through the development of new business models.

31. Imposing additional requirements on the renovation of multi-apartment buildings on the basis of the energy performance class of buildings could overcome the important regulatory barriers to reno-vation. However, the most important issue is to solve the problem of collective decision-making, taking into account the different demo-graphic profiles and preferences of apartment owners.

32. The Energy Service Company (ESCO) model may be modified and adapted for the renovation of multi-apartment buildings in Lithuania, following the example of the UK. Higher utility bills can be broken down according to the income of households living in apartment buildings requiring energy renovation. As in the case of subsidies for electricity, water and other utilities, life-line heat tariffs may be applied to pay for ESCO services. This would allow the cost of energy renovation to be shared between apartment owners with different incomes and address the principle of social justice; how-ever a case study conducted in Lithuania indicated a very low will-ingness to share energy renovation costs of multi-flat buildings with low-income neighbours.

33. Therefore, the implementation of life-line heat tariffs and the ESCO model for the promotion of energy renovation of multi-flat buildings is not a straightforward process and more case studies and investigations are necessary to develop and implement this scheme in Lithuania.

34. An assessment of WTP for energy renovation in multi-apartment buildings in Lithuania allowed the reparation of a concrete proposal for the application of the ESCO model can include on-billing financing model or financing based on monthly bills for heat consumption. While residents are reluctant to share renovation costs with lower-income neighbours, this innovative scheme could be experimentally applied in selected multi-apartment buildings to test its feasibility.

References

Abrahamse, W., Steg, L., Vlek, C. and Rothengatter, T. 2005. A Review of Intervention Studies Aimed at Household Energy Conservation. *Journal of Environmental Psychology* 25(3): 273–291.

Acito, F. & Jain, A. K. 1980. Evaluation of conjoint analysis results: A comparison of methods. *Journal of Marketing Research* 17(1): 106–112.

Adamowicz, W., Louviere, J. & Williams, M. 1994. Combining revealed and stated preference methods for valuing environmental amenities. *Journal of Environmental Economics and Management* 26(3): 271–292.

Akcura, E. 2015. Mandatory versus voluntary payment for green electricity. *Ecological Economics* 116(C): 84–94.

Alanne, K. & Saari, A. 2004. Sustainable small-scale CHP technologies for buildings: The basis for multi-perspective decision-making. *Renewable and Sustainable Energy Reviews* 8(5): 401–431.

Allen, S., Hammond, G. & McManus, M. C. 2008. Prospects for and barriers to domestic micro-generation: A United Kingdom perspective. *Applied Energy* 85(6): 528–544.

Aravena, C., Hutchinson, W. G. & Longo, A. 2012. Environmental pricing of externalities from different sources of electricity generation in Chile. *Energy Economics* 34(4): 1214–1225.

Asadi, E., Gameiro da Silva, M., Henggeler Antunes,C., Dias, L. 2012. Multi-objective optimization for building retrofit strategies: A model and an application. *Energy and Buildings* 44: 81–87.

Banfi, S., Farsi, M., Filippini, M. & Jakob, M. 2008. Willingness to pay for energy-saving measures in residential buildings. *Energy Economics* 30(2): 503–516.

Bator, F., M. 1958. The anatomy of market failure. *The Quarterly Journal of Economics* 72(3): 351–379, https://doi.org/10.2307/1882231

Beillan, V., Battaglini, E., Goater, A., Huber, A., Mayer, I., Trotignon, R. 2011. Barriers and drivers to energy efficient renovation in the residential sector Empirical findings from five European countries. *European Council for an Energy Efficient Economy*. Summer Study. Hyeres, France, 1083–1093.

Bell, C. J., Nadel, S., & Hayes, S. 2011. On-bill financing for energy efficiency improvements. *A review of current program challenges, opportunities and bets practices* (Report Number E118). Available online: www.aceee.org/sites/default/files/publications/researchreports/e118.pdf.

Bennett, J. & Blamey, R. 2001. *The choice modelling approach to environmental valuation*. Cheltenham, UK: Edward Elgar Publishing, 288 p.

Berrens, R. P., Bohara, A. K., Jenkins-Smith, H. C., Silva, C. L. & Weimer, D. L. 2004. Information and effort in contingent valuation surveys: Application to global climate change using national internet samples. *Journal of Environmental Economics and Management* 47(2): 331–363. Available online: https://doi.org/10.1016/S0095-0696(03)00094-9.

Bergmann, A., Colombo, S. & Hanley, N. 2008. Rural versus urban preferences for renewable energy developments. *Ecological Economics* 65(3): 616–625.

Bergmann, A., Hanley, N. & Wright, R. 2006. Valuing the attributes of renewable energy investments. *Energy Policy* 34(9): 1004–1014.

Berk, R., Fovell, R. G. 1999. Public Perceptions of Climate Change: A "Willingness to Pay Assessment". *Climatic Change* 41: 413–446.

Bertoldi, P., Rezessy, S., & Vine, E. 2006. Energy service companies in European countries: Current status and a strategy to foster their development. *Energy Policy* 34(14): 1818–1832.

Better Buildings Partnership 2010. Low carbon retrofit toolkit: A roadmap to success. London: The Building Centre. Available online: www.betterbuildingspartnetship. co.uk/sites/default/files/media/attachment/bbp-low-carbon-retrofit-toolkit. pdf.

Bigerna, S. & Polinori, P. 2014. Italian households' willingness to pay for green electricity. *Renewable and Sustainable Energy Reviews* 34: 110–121.

Boardman, B. 2010. *Fixing fuel poverty: Challenges and solutions*. London: Earthscan.

Bohringer, Ch., Vogt, C. 2004. The Dismantling of a Breakthrough: The Kyoto Protocol – Just Symbolic Policy! ZEW Discussion Paper No. 02-25. *The Canadian Journal of Economics* 36(2): 475–494.

Boie, I., Fernandes, C., Frías, P., & Klobasa, M. 2014. Efficient strategies for the integration of renewable energy into future energy infrastructures in Europe-An analysis based on transnational modeling and case studies for none European regions. *Energy Policy* 67: 170–185. Available online: https://doi: 10.1016/ j.enpol.2013.11.014.

Bollino, C. A. 2009. The willingness to pay for renewable energy sources: The case of Italy with socio-demographic determinants. *Energy Journal* 30(2): 81–96.

Borchers, A. M., Duke, J. M. & Parsons, G. R. 2007. Does willingness to pay for green energy differ by source? *Energy Policy* 35(6): 3327–3334. .

Boxall, P. C., Adamowicz, W. L., Swait, J., Williams, M. & Louviere, J. 1996. A comparison of stated preference methods for environmental valuation. *Ecological Economics* 18(3): 243–253.

Boute, A. 2012. Modernizing the Russian district heating sector: Financing energy efficiency and renewable energy investments under the New Federal Heat Law. *Pace Environmental Law Review* 29: 746–810.

Bouzarovski, S. 2014. Energy poverty in the European Union: Landscapes of vulnerability. *WIREs Energy and Environment* 3: 276–289. doi: 10.1002/wene.89.

Bouzarovski, S., & Petrova, S. 2015. A global perspective on domestic energy deprivation: Overcoming the energy poverty-fuel poverty binary.*Energy Research & Social Science* 10: 31–40.

BPIE. 2012. Nolte I, Griffiths N, Rapf O, Potcoava A. 2012. Implementing nearly zero-energy buildings (nZEB) in Romania – toward a definition and roadmap. Available online: http://bpie.eu/wp-content/uploads/2015/10/nZEB-Full-Report-Romania.pdf

Brandon, Gwendolyn and Lewis, Alan 1999. Reducing Household Energy Consumption: A Qualitative and Quantitative Field Study. *Journal of Environmental Psychology* 19: 75–85.

Brouwer, R., Brander L. and Van Beukering, P. 2008. A convenient truth: air travel passengers' willingness to pay to offset their CO2 emissions. *Climatic Change* 90(3): 299–313.

Browne, O., Poletti, S., & Young, D. 2015. How does market power affect the impact of large scale wind investment in 'energy only' wholesale electricity markets? *Energy Policy* 87: 17–27. Prieiga prie interneto: https://doi: 10.1016/ j.enpol.2015.08.030.

Brulez, D., Rauch, R. 1998. Energy Conservation Legislation in Thailand: Concepts, Procedures and Challenges. *Compendium on Energy Conservation Legislation in Countries of the Asia and Pacific Region.* UNESCAP. Bankok, Thailand.

Byrnes, L., Brown, C., Foster, J., & Wagner L. D. 2013. Australian renewable energy policy: Barriers and challenges. *Renewable Energy* 60(1): 711-721. Available online: https://doi: 10.1016/j.renene.2013.06.024.

Burger, V. 2013. The assessment of the regulatory and support framework for domestic buildings in Germany from the perspective of long-term climate protection targets. *Energy Policy* 59: 71–81.

Cadima, P. S. P. 2009. Retrofitting Homes for Better Energy Performance: e occupants' perspective. *PLAE 2009 – 26th Conference on Passive and Low- Energy Architecture,* Quebec City, Canada, 22–24 June 2009.

Cadoret, I., & Padovano, F. 2016. The political drivers of renewable energies policies. *Energy Economics* 56: 261–269. Available online: https://doi: 10.1016/j.eneco.2016.03.003.

Cameron, T. 2005. Individual option prices for climate change mitigation. *Journal of Public Economics* 89: 283–301.

Campbell, D., Hutchinson, W. G. & Scarpa, R. 2008. Incorporating discontinuous preferences into the analysis of discrete choice experiments. *Environmental and Resource Economics* 41(3): 401–417.

Capozza, M. 2006. Market mechanisms for white certificates trading, *Task XIV final report. OECD/IEA,* Paris. Carbon Monitoring for Action (CARMA). Available online: http://greencodeproject.org/en/carrna-power-plant-emissions.

Chan, K., Oerlemans, L. A. & Volschenk, J. 2015. On the construct validity of measures of willingness to pay for green electricity: Evidence from a south African case. *Applied Energy* 160: 321–328.

Chang, M. C., & Shieh, H. S. 2017. The Relations between Energy Efficiency and GDP in the Baltic Sea Region and Non-Baltic Sea Region. *Transformations in Business and Economics* 16(2): 235–247.

Chen, J., Cheng, S., Vinko, N., & Song, M. 2018. Quo Vadis? Major players in global coal consumption and emissions reduction. *Transformations in Business and Economics* 17(1): 112–132.

Chen, J., Huang, J., Zheng, L., & Zhang, C. 2019. An Empirical Analysis of Telecommunication Infrastructure Promoting the Scale of International Service Trade: Based on the Panel Data of Countries along the Belt and Road. *Transformations in Business and Economics* 18(2): 124–139.

Cho, C., Yang, L., Chu, Y., & Yang, H. 2013. Renewable energy and renewable R&D in EU countries: a cointegration analysis. *Asian Journal of Natural and Applied Sciences* 2(1): 10–16.

Christensen, T. H., Gram-Hanssen, K., Adjei, A., & de Best-Waldhober, M. 2011. Energy renovation practices in Danish homes: The influence of energy labels on home renovation practices. Paper presented at 10th Conference of the European Sociological Association (esa2011): *Social relations in turbulent times,* Geneva, Switzerland. Available online: https://vbn.aau.dk/en/publications/energy-renovation-practices-in-danish-homes-the-influence-of-ener.

Christiaensen, L. J. & Sarris, A. 2007. Rural household vulnerability and insurance against commodity risks: Evidence from the United Republic of Tanzania. *FAO Commodities and Trade Technical Paper.* Available online: www.fao.org/3/a1310e/a1310e.pdf.

Claudy, M., O'Driscoll, A. 2008. Beyond Economics – A behavioural approach to energy efficiency in domestic buildings. *Euro-Asian Journal of Sustainable Energy Development Policy* 1: 27–40.

Claudy, M. C. Michelsen, C. & O'Driscoll, A. 2011. The diffusion of microgeneration technologies–assessing the influence of perceived product characteristics on home owners' willingness to pay. *Energy Policy* 39(3): 1459–1469.

Claudy, M. C., Michelsen, C., O'Driscoll, A. & Mullen, M. R. 2010. Consumer awareness in the adoption of microgeneration technologies: An empirical investigation in the republic of Ireland. *Renewable and Sustainable Energy Reviews* 14(7): 2154–2160.

Collins M. & Curtis J. 2018. Willingness-to-pay and free-riding in a national energy efficiency retrofit grant scheme. *Energy Policy* 118: 211–220.

Conner, Helene 1999. Taking Non-monetizable Impacts (NMIs) into Account in an Eco Development Strategy. *Valuation and the Environment*. Northampton, Massachusetts: Edward Elgar Publishing, 241–262.

Consumer Focus 2012. What's in it for me? Using the benefits of energy efficiency to overcome the barriers. Energy Efficiency Partnership for Buildings (EEPB). *Breaking Barriers. Summary Report*, London.

Crilly, M. Lemon, A.J. Wright, M.B. Cook, D. Shaw 2012. Retrofitting homes for energy efficiency: an integrated approach to innovation in the low-carbon overhaul of UK social housing. *Energy & Environment* 23: 1027–1056.

Crossley, D., Maloney, M., Watt, G. 2000. Developing mechanisms for promoting demand-side management and energy efficiency in changing electricity businesses. *Task VI of the IEA Demand-Side Management Program*. Hornsby Heights, Australia. Available online: http://dsm.iae.org/NewDSM/Prog/Library/upload/28/resrpt3_fin.pdf.

Dabija, A.-M. 2010. Rehabilitation of Mass Dwellings in Romania. *A Critical Approach. Scientific Bulletin of the Electrical Engineering Faculty* 10(3): 40–45.

Dagher, L. & Harajli, H. 2015. Willingness to pay for green power in an unreliable electricity sector: Part 1. The case of the Lebanese residential sector. *Renewable and Sustainable Energy Reviews*, 50(C): 1634–1642.

De Groot, R. S., Wilson, M. A. & Boumans, R. M. 2002. A typology for the classification, description and valuation of ecosystem functions, goods and services. *Ecological Economics*, 41(3): 393–408.

Deci, E., Koestner, R., and Ryan, R. M. 1999. A meta-analytic review of experiments examining the effects of extrinsic rewards on intrinsic motivation. *Psychological Bulletin* 125: 627–668.

De Paepe, M., D'Herdt, P. & Mertens, D. 2006. Micro-CHP systems for residential applications. *Energy Conversion and Management*, 47(18): 3435–3446.

De T'Serclaes, P. 2007. *Financing energy efficient homes. Existing policy responses to financial barriers*. Paris: IEA Information paper.

De T'Serclaes, P. & Jollands, N. 2007. *Mind the Gap. Quantifying Principal-Agent Problems in Energy Efficiency*. Paris: IEA Information paper.

Dias, R., Mattos, C., Balesieri, J. 2004. Energy education: breaking up the rational energy use barriers. *Energy Policy* 32 (11):1339–1347.

Edomah, N., Foulds, C., & Jones, A. 2017. Influences on energy supply infrastructure: A comparison of different theoretical perspectives. *Renewable Sustainable Energy Reviews* 79: 765–778. Available online: https://doi: 10.1016/j.rser.2017.05.072.

EFA (Energy Futures Australia) 2002. Mechanisms for promoting societal demand management. *Independent Pricing and Regulatory Tribunal (IPART) of New South Wales.* Research paper 19, Sydney.

Ek, K. 2005. Public and private attitudes towards "green" electricity: The case of Swedish wind power. *Energy Policy* 33(13): 1677–1689.

EST (Energy Saving Trust) 2010. *Sustainable refurbishment: Towards an 80 per cent reduction in CO2 emission, water efficiency, waste reducion, and climate change adaptation.* London.

ESTEP 2019a. Financing of Lithuanian economic sectors after 2020 assessment: Energy efficiency and housing renovation. Annex 7 to the Final Report. Vilnius. https://lrv.lt/uploads/main/documents/files/Energijos%20efektyvumas%20ir%20b%C5%ABsto%20renovacija(1).pdf

ESTEP 2019b. Evaluation of financing of Lithuanian economic sectors after 2020. Final report. Vilnius, 34 p. Available at: https://www.esinvesticijos.lt/media/force_download/?url=/uploads/main/documents/docs/107931_f72468970def8959587f7d51cbad9785.pdf

European Commission. 2012. DIRECTIVE 2012/27/EU OF THE EUROPEAN PARLIAMENT AND OF THE COUNCIL of 25 October 2012 on energy efficiency, amending Directives 2009/125/EC and 2010/30/EU and repealing Directives 2004/8/EC and 2006/32/EC available at: https://eur-lex.europa.eu/LexUriServ/LexUriServ.do?uri=OJ:L:2012:315:0001:0056:en:PDF

Evander, A., G. Siebock, and L. Neij 2004. Diffusion and development of new energy technologies: lessons learned in view of renewable energy and energy efficiency end-use projects in developing countries Lund, Sweden: International Institute for Industrial Environmental Economics. *Report 2004: 2.*

Eyre, N. 1998. A golden age or a false dawn? Energy efficiency in UK competitive energy markets. *Energy Policy* 26(12): 963–972.

European Commission 2019. European Construction Sector Observatory. Policy fact sheet. Lithuania. Multi-apartment Renovation Programme. Thematic Objective 1. Available online: http://ECSO_PFS_LT_MARP_2019.pdf.

Fawcett, T., Killip, G., Janda, K. 2013. Building expertise: identifying policy gaps and new ideas in housing eco-renovation in the UK and France. *Proceedings of ECEEE 2013,* Summer Study. Belambra Les Criques, France, 339–350.

Frederiks, E.R., Stenner K., Hobman, E.V. 2015. Household energy use: Applying behavioural economics to understand consumer decision-making and behaviour. *Renewable and Sustainable Energy Reviews* 41: 1385–1394.

Frey, B. 1999. Morality and rationality in environmental policy. *Journal of Consumer Policy* 22(4): 395–417.

Frey, B. and Jegen, R. 2001. Motivation crowding theory: a survey of empirical evidence. *Journal of Economic Surveys* 15: 589–611.

Frey, B. 2005. Excise taxes: Economics, politics, and psychology. *Theory and Practice of Excise Taxation: Smoking, Drinking, Gambling, Polluting, and Driving* [ed. S. Cnossen]. Oxford, UK: Oxford University Pres.

Friege, J.,Chappin, E. 2014. Modelling decisions on energy-efficient renovations: A riew. *Renewable and Sustainable Energy Reviews* 39: 196–208.

Galvin, R. 2015 Integrating the rebound effect: Accurate predictors for upgrading domestic heating. *Building Research & Information* 43(6): 710–722.

Garcia, R., Bardhi, F. & Friedrich, C. 2007. Overcoming consumer resistance to innovation. *MIT Sloan Management Review* 48(4): 82.

Geller, H., P. Harrington, P., Rosenfeld, A.H.,Tanishima S., Unander, F. 2006. Policies for increasing energy efficiencies. 30 years of experience in OECD countries. *Energy Policy* 34(5): 556–573.

Gilligham, K., Newell, R., Palmer, K. 2004. The Effectiveness and Cost of Energy Efficiency Programmes, Resources, Resources for the Future. *Technical Paper*.

Gifford, Robert. 2011. The Dragons of Inaction: Psychological Barriers That Limit Climate Change Mitigation and Adaptation. *American Psychologist* 66(4): 290–302. Available online: http://10.1037/a0023566.

Gifford, R., Kormos, Ch., McIntyre, A. 2011. Behavioral dimensions of climate change: drivers, responses, barriers, and interventions. *WIREs Climate Change* 2: 801–827. Available online: http://doi.10.1002/wcc.143.

Goodin, R. 1994. Selling environmental indulgences. *Kyklos* 47: 573–96. Available online: https://doi.org/10.1111/j.1467-6435.1994.tb02067.x.

Golove, W.H., Eto, J. H. 1996. *Market barriers to energy efficiency: A critical reappraisal of the rationale for public policies to promote energy efficiency*. Lawrence Berkeley National Laboratory, University of California, Berkeley, California. Available online: https://emp.lbl.gov/publications/market-barriers-energy-efficiency.

Grilli, G., Balest, J., Garegnani, G. & Paletto, A. 2015. Exploring residents' willingness to pay for renewable energy supply: Evidences from an Italian case study. *Journal of Environmental Accounting and Management* 4(2): 105–113.

Grosche P. & Vance C. 2009. Willingness-to-Pay for Energy Conservation and Free-Ridership on Subsidization – Evidence from Germany. *Energy Journal* 30(2):135–154.

Guerra-Santin, O., Boess, S., Konsttantinou, T., Romero-Herrera, N. 2017. Designing for residents: building monitoring and co-creation in social housing renovation in the Netherlands. *Energy Research & Social Science* 32: 164–179.

Gullberg, A. T., Ohlhorst, D., & Schreurs, M. 2014. Towards a low carbon energy future-Renewable energy cooperation between Germany and Norway. *Renewable Energy* 68: 216–222. Available online: https://doi: 10.1016/j.renene.2014.02.001.

Guo, X., Liu, H., Mao, X., Jin, J., Chen, D. & Cheng, S. 2014. Willingness to pay for renewable electricity: A contingent valuation study in Beijing, China. *Energy Policy* 68: 340–347.

Hanley, N., Colombo, S., Kriström, B. & Watson, F. 2009. Accounting for negative, zero and positive willingness to pay for landscape change in a national park. *Journal of Agricultural Economic,* 60(1): 1–16.

Hanley, N. & Nevin, C. 1999. Appraising renewable energy developments in remote communities: The case of the North Assynt estate, Scotland. *Energy Policy* 27(9): 527–547.

Harrison, B. 2015. Expanding the renewable energy industry through tax subsidies using the structure and rationale of traditional energy tax subsidies. *University of Michigan Journal of Law Reform* 48(3): 845–877.

Healy, J.D.2004. *Housing, fuel poverty, and health: a pan-European analysis*. Ashgate Publishing, New York.

Hein Nybroe, M. 2001. DSM in Denmark after liberalization. *Proceedings of Summer Study on Energy in Buildings*. Stockholm: ECEEE. Paris ADEME Editions.

Heiskanen et al. 2012. *Working paper*: Literature review of key stakeholders, users and investors D2.4. of WP2 of the Entranze Project. Available online: www.entranze.eu/files/downloads/D2_4/D2_4_Complete_FINAL3.pdf.

Hensher, D. A. 2010. Hypothetical bias, choice experiments and willingness to pay. *Transportation Research Part B: Methodological* 44(6): 735–752.

Herbes, C., Friege, C., Baldo, D. & Mueller, K. 2015. Willingness to pay lip service? Applying a neuroscience-based method to WTP for green electricity. *Energy Policy* 87: 562–572.

Herfert, G., & Lentz, S. 2007. New spatial patterns of population development as a factor in restructuring Eastern Germany. *Restructuring Eastern Germany* [ed., S. Lentz]. Wiesbaden: Springer, 97–109. Available online: https://doi:10.1007/978-3-540-32088-3-7.

Hernández, D., & Bird, S. 2010. Energy burden and the need for integrated low-income housing and energy policy. *Poverty & Public Policy* 2(4): 668–688.

Herrero, S. T., Urge-Vorsatz, D. 2012. Trapped in the heat: a post-communist type of fuel poverty. *Energy Policy* 49: 60–68.

Hole, A. R. 2007. Estimating mixed logit models using maximum simulated likelihood. *Stata Journal* 7(3): 388–401.

Hori, S., Kondo, K., Nogata, D., Ben, H. 2013. The determinants of household energy-saving behavior: Survey and comparison in five major Asian cities. *Energy Policy* 52: 354–362.

Horne, R., Dalton, T. 2014. Transition to low carbon? *An analysis of socio-technical change in housing renovation.* SAGE publishing, Australia. Available online: https://doi.org/10.1177/0042098013516684.

Hoyos, D., Longo, A., Markandya, A. 2009. WTP for Global and Ancillary Benefits of Climate Change Mitigation: Preliminary Results. *17th Annual Conference of the European Association of Environmental and Resource Economists (EAERE).* University of Bath, Amsterdam, 24–27 June.

Huang, S., Lo, S., & Lin, Y. 2013 To re-explore the causality between barriers to renewable energy development: A case study of wind energy. *Energies* 6(9): 4465–4488. Available online: https://doi: 10.3390/en6094465.

Huber, A., Mayer, I., Beillan, V., Goater, A., Trotignon, R. & Battalgini E. 2011. Refurbishing residential buildings. *A socio-economic analysis of retrofitting projects in five countries.* Available online: http://fedarene.org/documents/pro- jects/EEW2/WSED2011/Huber.pdf.

IEA 2008 Promoting Energy Efficiency Investments. *Case Studies from the Residential Sector.* Paris: IEA, 324 p. Available online: https://doi.org/10.1787/9789264042155-en.

IIBW 2008. Implementation of European Standards in Romanian Housing Legislation. *Final Report.* Available online: www.rabo.org.ro/wp-content/uploads/RHL-Final- Report-080129-en.pdf.

Itard, L., Meijer, M, Vrins, E & Hoiting, H. 2008. Building Renovation and Modernisation in Europe: State of the art review. *ERABuild Final Report.* Delft, Netherlands, 220 p.

Jacobsen G., Kotchen, M. and Vanderbergh, M. 2012. The behavioural response to voluntary provision of an environmental public good: evidence from residential electricity demand. *European Economics Review* 56: 946–960. Available online: https://<doi:10.1016/j.euroecorev.2012.02.008.

James, J.S., Rickard, B.J. & Rossman, W.J. 2009. Product differentiation and market segmentation in applesauce: Using a choice experiment to assess the value of organic, local, and nutrition attributes. *Agricultural & Resource Economics Review* 38(3): 357.

Janda, K. B., Parag, Y. 2013. A middle-out approach for improving energy performance in buildings *Building Research and Information* 41: 39–50. Available online: https://doi.org/10.1080/09613218.2013.743396.

Janda, K. B., G. Killip, G. Fawcett, T. 2014. Reducing Carbon from the "Middle-Out": The Role of Builders in Domestic Refurbishment. *Buildings* 4(4): 911–936.

Janssen, R. 2004. Towards Energy Efficient Buildings in Europe. *Final Report.* London. Available online: http://ec.europa.eu/energy/library/en_bat_en.pdf.

Jeeninga, H., Uyterlinde, M.A. 2000. The sky is the limit! Or why can more efficient appliances not decrease the electricity consumption of Dutch households. *Energy efficiency in household appliances and lighting* [eds. P. Bertoldi, A. Ricci, A. Almeida]. Heidelberg: Springer, 806–817.

Jensen, P. A., Malsea, E. 2015. Value based building renovation – a tool for decision-making and evaluation. *Building and Environment* 92: 1–9.

Jensen, S., Mohlin, K., Pittel, K., & Sterner, T. 2015. An introduction to the green paradox: the unintended consequences of climate policies. *Review of Environmental Economics and Policy* 9(2): 246–265. Available online: https://doi: 10.1093/reep/rev010.

Johnson, E., Nemet, G. F. & Nemet, G. 2010. Willingness to pay for climate policy: A review of estimates. *Working Paper Series. La Follette School Working Paper* 2010–2011. Available online: www.lafollette.wisc.edu/publications/workingpapers.

Johnson, K., Willoughby, G., Shimoda, W., & Volker, M. 2012. Lessons learned from the field: Key strategies for implementing successful on-the-bill financing programs. *Energy Efficiency* 5: 109–119.

Jovovic, R., Simanaviciene, Z., & Dirma, V. 2017. Assessment of Heat Production Savings Resulting from Replacement of Gas with Biofuels. *Transformations in Business & Economics* 16(1)(40): 34–51.

Jung, W., Kim, T. & Lee, S. T. 2015. The study on the value of new & renewable energy as a future alternative energy source in Korea.

Kahneman and Tversky, A. 1979. Prospect theory: an analysis of decision under risk. *Econometrica* 47(2): 263–291.

Karakaya, E., & Sriwannawit, P. 2015. Barriers to the adoption of photovoltaic systems: The state of the art. *Renewable Sustainable Energy Reviews* 49: 60–66. Available online: https://doi: 10.1016/j.rser.2015.04.058.

Karatayev, M., Hall, S., Kalyuzhnova, Y., & Clarke, M. L. 2016. Renewable energy technology uptake in Kazakhstan: Policy drivers and barriers in a transitional economy. *Renewable Sustainable Energy Review* 66: 120–136. Prieiga per injternetą: https://doi: 10.1016/j.rser.2016.07.057.

Karvonen, A. 2013. Towards systemic domestic retrofit: A social practices approach. *Building Research & Information* 41(5): 563–574.

Kilinc-Ata, N. 2016. The evaluation of renewable energy policies across EU countries and US states: An econometric approach. *Energy for Sustainable Development* 31: 83–90. Available online: https://doi: 10.1016/j.esd.2015.12.006

Killip, G. 2013. Products, practices and processes: Exploring the innovation potential for low-carbon housing refurbishment among small and medium-sized enterprises (SMEs) in the UK construction industry". *Energy Policy* 62: 522–530.

Killip, G., Fawcett, T., Janda, K. 2014. Innovation in low-energy residential renovation: UK and France. *Proceedings of ICE: Energy* 167(3): 117–124.

Kosenius, A. & Ollikainen, M. 2013. Valuation of environmental and societal trade-offs of renewable energy sources. *Energy Policy* 62: 1148–1156.

Kraeusel, J. & Möst, D. 2012. Carbon capture and storage on its way to large-scale deployment: Social acceptance and willingness to pay in Germany. *Energy Policy* 49: 642–651.

Krugman, P. and Wells, R. 2006. *Economics*, New York, Worth Publishers, 864 p.

Kushler, M., Work, D., Witte, P. 2004. Five years in: an examination of the first half decade of public benefits energy efficiency policy. American Council for an Energy Efficient Economy. *Report U041.*

abanca N., Suerkemper, F., Bertoldi, P., Irrek, W., Duplessis, B. 2015. Energy efficiency services for residential buildings: market situation and existing potentials in the European Union. *Journal of Cleaner Production* 109: 284–295.

Lancaster, K. J. 1966. A new approach to consumer theory. *Journal of Political Economy* (74):132–157.

Lee, C. & Heo, H. 2016. Estimating willingness to pay for renewable energy in South Korea using the contingent valuation method. *Energy Policy* 94: 150–156.

Leung B. C.-M. 2018. Greening existing buildings [GEB] strategies. *Energy Reports* 4: 159–206.

Levine, M., Urge-Vorsatz, D., Blok, K., Geng, L, Harvey, D., Land, S., Levermore, G., Mongameli Mehlwana, A., Mirasgedis, S., Novikova, A., Riling, J., Yoshino, H. 2007, *Residential and commercial buildings, Climate Change 2007: Mitigation, Contribution of Working Group III to the Fourth Assessment Report of the Intergovernmental Panel on Climate Change* [eds. B. Metz, O.R. Davidson, PR. Bosch, R. Dave, L.A. Meyer]. Cambridge University Press: Cambridge, U.K. & New York, U.S.A.

Ministry of Environment of the Republic of Lithuania. 2012.

The National Strategy for Climate Change Management Policy of Lithuania.

Available at: www.preventionweb.net/files/58659_nacionalineklimatokaitosvaldym opoli.pdf.

Ministry of Environment of the Republic of Lithuania. 2020. National Energy and Climate Action Plan of the Republic of Lithuania 2021–2030, Vilnius. Available at: https://ec.europa.eu/energy/sites/ener/files/documents/lt_final_necp_main_en.pdf.

Ministry of Energy of the Republic of Lithuania. 2018. National Energy Independence Strategy, Vilnius, Available online: https://enmin.lrv.lt/uploads/enmin/documents/files/Nacionaline%20energetines%20nepriklausomybes%20 strategija_2018_EN.pdf.

Government of the Republic of Lithuania. 2012. The National Progress Programme 2014–2020 Available online: https://e-seimas.lrs.lt/portal/legalAct/lt/TAD/TAIS.439028.

Lin, J. 2002. *Made for China: Energy efficiency standards and labels for household appliances.* Berkeley, California, 11 p.

Liu N., Zhao Y., Ge J. 2018. Do renters skimp on energy efficiency during economic recessions? Evidence from Northeast Scotland. *Energy* 165: 164–75.

Longo A., Hoyos D., Markandya A. 2012. Willingness to Pay for Ancillary Benefits of Climate Change Mitigation. *Environment Resource Economy* 51: 119–140.

Longo, A., Markandya, A. & Petrucci, M. 2008. The internalization of externalities in the production of electricity: Willingness to pay for the attributes of a policy for renewable energy. *Ecological Economics* 67(1): 140–152.

Louviere, J. J., Flynn, T. N. & Carson, R. T. 2010. Discrete choice experiments are not conjoint analysis. *Journal of Choice Modelling* 3(3): 57–72.

Louviere, J. J. & Hensher, D. A. 1982. Design and analysis of simulated choice or allocation experiments in travel choice modeling. *Journal of Marketing Research* 20(4): 350–367.

Lu J, Ren L, Yao S, Rong D, Skare M, Streimikis J. 2020. Renewable energy barriers and coping strategies: Evidence from the Baltic States. *Sustainable Development* 28: 352–367. Online ISSN:1099-1719. Available online: https://doi.org/10.1002/sd.2030.

Lujanen, M. 2010 Legal challenges in ensuring regular maintenance and repairs of owner-occupied apartment blocks, *International Journal of Law in the Built Environment* 2: 178–197.

Lutzenhiser, Loren 1993. Social and Behavioral Aspects of Energy Use. *Annual Review of Energy Environment* 18: 247–289.

Lyu, X., & Shi, A. 2018. Research on the Renewable Energy Industry Financing Efficiency Assessment and Mode Selection. *Sustainability* 10(1): 222. Available online: https://doi: 10.3390/su10010222.

Ma, Z., Cooper, P., Daly, D., Ledo, L. 2012. Existing building retrofits: Methodology and state of the art. *Energy and Buildings* 55: 889–902.

Ma Ch., Rogers A. A., Kragt M. E., Zhang F., Polyakov M., Gibson F., Chalak M., Pandit R., Tapsuwan S. 2015. Consumers' Willingness to pay for renewable energy: A meta-regression analysis. *Resource and Energy Economics* 42: 93–109. Available online: https://doi: 10.1016/j.reseneeco.2015.07.003.

Malik, K., Rahman, S. M., Khondaker, A. N., Abubakar, I. R., Aina, Y. A., & Hasan, M. A. 2019. Renewable energy utilization to promote sustainability in GCC countries: policies, drivers, and barriers. *Environmental Science and Pollution Research* 26(20): 20798–20814. Available online: https://doi: 10.1007/s11356-019-06138-2.

Mallaband, B., Haines,V., & Mitchell,V. 2012. Barriers to domestic retrofit – learning from past home improvement experiences. Retrofit, Salford, Manchester. Available online: www.energy.salford.ac.uk/retrofit-salford-2012.

Markandya, A., & Steimikiene, D. 2003. Efficiency and Affordability Considerations In The Pricing Of Energy For Households. *Economic Journal of Development Issues* 3(2): 1–14.

Matschoss, K., Heiskanen, E., Atanasiu, B., Kranzl, L. 2013. Energy renovation of EU multifamily buildings: do current policies target the real problems? *Proceedings of ECEEE 2013*. Summer Study, Belambra Les Criques, France, 1485–1496.

McFadden, D. & Train, K. 2000. Mixed MNL models for discrete response. *Journal of Applied Econometrics* 15: 447–470.

Medema, S. G. 2007. The Hesitant Hand: Mill, Sidgwick, and the Evolution of the Theory of Market Failure. *History of Political Economy* 39(3): 331–58.

Menegaki, A. 2008. Valuation for renewable energy: A comparative review. *Renewable and Sustainable Energy Reviews* 12(9): 2422–2437.

Ministry of Energy of Republic of Lithuania. 2014. Energy Efficiency Action Plan, Vilnius, available at: https://ec.europa.eu/energy/sites/ener/files/documents/lt_neeap2014_annex_1_annual_report_2014_en.pdf

Moore, R. 2012. Definitions. of fuel poverty: Implications for policy. *Energy Policy* 49: 19–26.

Morita, T. & Managi, S. 2015) Consumers' willingness to pay for electricity after the great East Japan earthquake. *Economic Analysis and Policy* 48: 82–105.

Moskovitz D. 1993. Green pricing: Customer choice moves beyond IRP. *The Electricity Journal* 6(8): 42–50.

Nair, G., Gustavsson, L. & Mahapatra, K. 2010. Factors influencing energy efficiency investments in existing Swedish residential buildings. *Energy Policy* 38: 2956–2963.

Nasirov, S., Silva, C., & Agostini, C. A. 2015. Investors' perspectives on barriers to the deployment of renewable energy sources in Chile. *Energies* 8(5): 3794–3814. Available online: https://doi: 10.3390/en8053794.

Navrud, S. & Bråten, K. G. 2007. Consumers' preferences for green and brown electricity: A choice modelling approach. *Revue D'Économie Politique* 117(5): 795–811.

Nesta, L., Vona, F., & Nicolli, F. 2014. Environmental policies, competition and innovation in renewable energy. *Journal of Environmental Economics and Management* 67(3): 396–411. Available online: https://doi: 10.1016/j.jeem.2014.01.001.

Newell, R.G. & Siikamkj, J.V. 2013. Nudging energy efficiency behavior: the role of information labels. *Journal of the Association of Environmental and Resource Economics* 1: 555–598.

Nikola, N. 2011. The effect of pipe repairs on housing prices. Master's thesis, Aalto University, School of Economics, Department of Finance. Online: http://epub. lib.aalto.fi/fi/ ethesis/pdf/12524/hse_ethesis_12524.pdf.

Nomura, N. & Akai, M. 2004. Willingness to pay for green electricity in Japan as estimated through contingent valuation method. *Applied Energy* 78(4): 453–463.

Novikova, A., Vieider, F., Neuhoff, K., & Amecke, H. 2011. Drivers of thermal retrofit decisions: *A survey of German single- and two-family houses (CPI Report).* Berlin: Climate Policy Initiative Berlin.

Oberst, C. & Madlener, R. 2014. Prosumer preferences regarding the adoption of micro-generation technologies: Empirical evidence for German homeowners. Available online: https://econpapers.repec.org/paper/risfcnwpa/2014_5f022. htm.

O'Conner, Martin and Spash, Clive L. 1999. *Valuation and the environment: Theory, method and practise.* Northampton, Massachusetts: Edward Elgar Publishing, 352 p.

Ohunakin, O. S., Adaramola, M. S., Oyewola, O. M., & Fagbenle, R. O. 2014. Solar energy applications and development in Nigeria: Drivers and barriers. *Renewable and Sustainable Energy Reviews* 32: 294–301. Available online: https:// doi: 10.1016/j.rser.2014.01.014.

O'Neill, Brian C. and Belinda S. Chen 2002. Demographic Determinants of Household Energy Use in the United States. *Population and Development Review. Supplement: Population and Environment: Methods of Analysis* 28: 53–88.

Organ, S., Proverbs, D., Squires, G. 2013. Motivations for energy efficiency refurbishment in owner-occupied housing, *Structural Survey* 31: 101–120.

Owens S, Driffill L. 2008. How to change attitudes and behaviours in the context of energy. *Energy Policy* 36(12): 4412–4418.

Ozarisoy, B.; Altan, H. 2017. Adoption of Energy Design Strategies for Retrofitting Mass Housing Estates in Northern Cyprus. *Sustainability* 9: 1477.

Paiho, S., Abdurafikov, R., Hoang, H., Kuusisto 2015. An analysis of different business models for energy efficient renovation of residential districts in Russia cold regions. *Sustainable Cities and Societies* 14: 31–42.

Painuly, J. P. 2001. Barriers to renewable energy penetration; a framework for analysis. *Renewable Energy* 24(1): 73–89. Available online: https://doi: 10.1016/S0960-1481(00)00186-5.

Palmer, K. 1999. Electricity Restructuring: Shortcut or Detour on the Road to Achieving Greenhouse Gas Reductions? *Resources for the Future Climate Issue Brief 18*. Passive House Institute. Available online: www. passivehouse.com.

Paravantis, J. A., Stigka, E. K., & Mihalakakou, G. K. 2014. An analysis of public attitudes towards renewable energy in Western Greece. *Renewable and Sustainable Energy Reviews* 32: 100–106. Available online: https://doi: 10.1109/IISA.2014.6878776.

Ploeg, F., & Withagen, C. 2015. Global warming and the green paradox: a review of adverse effects of climate policies. *Review of Environmental Economics and Policy* 9(2): 285–303. Available online: https://doi: 10.1093/reep/rev008.

Pollitt, M.G., & Shaorshadze, I. 2011. The role of behavioural economics in energy and climate policy. *EPRG Working Paper* 1130. Cambridge Working Paper in Economics 1165.

Polzin, F., Migendt, M., Täube, F. A., & Von Flotow, P. 2015. Public policy influence on renewable energy investments-a panel data study across OECD countries. *Energy Policy* 80: 98–111. Available online: https://doi: 10.1016/j.enpol.2015.01.026.

Poortinga, W.; Steg, Vlek, Ch. 2004. Values, Environmental Concern, and Environmental Behavior: A Study into Household Energy Use. *Environment and Behavior* 36(1): 70–93.

Poortinga, W., Steg, L., Vlek, C. & Wiersma, G. 2003. Household preferences for energy-saving measures: A conjoint analysis. *Journal of Economic Psychology* 24(1): 49–64.

Portnov B. A., Trop T., Svechkina A., Ofek S., Akron S., Ghermandi A. 2018. Factors affecting homebuyers' willingness to pay green building price premium: Evidence from a nationwide survey in Israel. *Building and Environment* 137: 280–291.

Pothitou, M., Hanna, R.F., Chalvatzis, K.J. 2016. Environmental knowledge, pro-environmental behaviour and energy savings in households: An empirical study. *Applied Energy* 184: 1217–1229.

Risch, C. A. 2012. Evaluation of the impact of environmental public policy measures on energy consumption and greenhouse gas emissions in the French residential sector, *Energy Policy* 46: 170–184.

Ritchey, T. 2011. Wicked problems – social messes. *Decision support modelling with morphological analysis*. Berlin, Heidelberg: Springer. Available online: https://doi: 10.1007/978-3-642-19653-9.

Rittel, H. W. J., & Webber, M. M. 1973. Dilemmas in a general theory of planning. *Policy Sciences* 4: 155–169. Available online: https://doi: 10. 1007/BF01405730.

Qi, S., & Li, Y. 2017. Threshold effects of renewable energy consumption on economic growth under energy transformation. *Chinese Journal of Population Resources and Environment* 15(4): 312–321. Available online: https://doi: 10.1080/10042857.2017.1416049.

Ramos, A., Gago, A., Labandeira, X., Linares, P. 2015. The role of information for energy efficiency in the residential sector. *Energy Economics* 52: 517–529.

Raza, W., Saula, H., Islam, S. U., Ayub, M., Saleem, M., & Raza N. 2015. Renewable energy resources: Current status and barriers in their adaptation for Pakistan. *Journal of Bioprocessing and Chemical Engineering* 3(3): 1–9.

Reynolds, T., Kolodinsky, J., Murray, B. 2012. Consumer preferences and willingness to pay for compact fluorescent lighting: Policy implications for energy efficiency promotion in Saint Lucia. *Energy Policy* 41: 712–722.

Revelt, D. & Train, K. 1998. Mixed logit with repeated choices: Households' choices of appliance efficiency level. *Review of Economics and Statistics* 80(4): 647–657.

Ries, C. R, Jenkins, J., O. Wise, O. 2009. Improving the Energy Performance of Buildings: Learning from the European Union and Australia, *Technical Report*. Rand Corporation, 40 p. Available online: www.rand.org/pubs/technical_reports/TR728.html.

Risch, Ch. 2012. Evaluation of the impact of environmental public policy measures on energy consumption and greenhouse gas emissions in the French residential sector. *Energy Policy* 46: 170–184.

Roe, B., Teisl, M. F., Levy, A. & Russell, M. 2001. US consumers' willingness to pay for green electricity. *Energy Policy* 29(11): 917–925.

Rose, S., Clark, G., Poe, D., Rondeau, D. and Schulze, W. 2002. Field and laboratory tests of a provision point mechanism. *Resource and Energy Economics* 24 :131–155.

Samuelson, W., Zeckhauser, R. 1988. Status quo bias in decision making. *Journal of Risk and Uncertainty* 1: 7–59. https://doi.org/10.1007/BF00055564.

Sardianou, E. 2007. Estimating energy conservation patterns of Greek households. *Energy Policy* 35(7): 3778–3379.

Sardianou, E. & Genoudi, P. 2013. Which factors affect the willingness of consumers to adopt renewable energies? *Renewable Energy* 57: 1–4.

Sauter, R. & Watson, J. 2007. Strategies for the deployment of micro-generation: Implications for social acceptance. *Energy Policy* 35(5): 2770–2779.

Scarpa, R. & Willis, K. 2010. Willingness-to-pay for renewable energy: Primary and discretionary choice of British households' for micro-generation technologies. *Energy Economics* 32(1): 129–136.

Scott, S. 1997. Household energy efficiency in Ireland: A replication study of owner-ship of energy saving items. *Energy Economics* 19: 187–208.

Seetharaman, Moorthy, K., Patwa, N., Saravan, & Gupta, Y. 2019 Breaking barriers in deployment of renewable energy. *Heliyon* 5(1): 1–23. Available online: https://doi: 10.1016/j.heliyon.2019.e01166.

Shen, J., and Sajo, T. 2009. Does energy efficiency labours alter consumers' purchasing decisions? A latent class approach based on a stated choice experiment in Shanghai. *Journal of Environmental Management* 90(11): 3561–3573.

Shogren, F. and Taylor, L. 2008. On behavioral-environmental economics. *Review of Environmental Economics and Policy* 2(1): 26–44.

Sinn, H. W. 2015. Introductory comment-the green paradox: a supply-side view of the climate problem. *Review of Environmental Economics and Policy* 9(2): 239–245. Available online: https://doi: 10.1093/reep/rev011.

Sirombo, E., Filippi, M., Catalano A. 2017. Building monitoring system in a large social housing intervention in Northern Italy. *Energy Procedia* 140: 389–397.

Solomon, B. D., Johnson, N. H. 2009. Valuing Climate Protection through Willingness to pay for Biomass Ethanol. *Ecological Economics* 68: 2137–2144.

Sorrell, S., O'Malley, E., Schleich, J. & Scott, S. 2004. The Economics of Energy Efficiency – Barriers to Cost-Effective Investment. Cheltenham: Edward Elgar, 349 p.

Sovacool, B. K. 2009. Rejecting renewables: The socio-technical impediments to renewable electricity in the United States. *Energy Policy* 37(11): 4500–4513. Available online: https://doi: 10.1016/j.enpol.2009.05.073.

Sovacool, B. K., & Saunders, H. 2014. Competing policy packages and the complexity of energy security. *Energy* 67: 641–651. Available online: https://doi: 10.1016/j.energy.2014.01.039.

Skema, R., Dzenajaviciene, F. 2017. [Lithuania]. Renovation Programme with EU funding. *Case study* prepared by LEI for EPATEE project, funded by European Union's Horizon 2020 programme. Available online: https://epatee.eu/system/tdf/epatee_case_study_lithuania_renovation_programme_with_eu_funding_ok_0.pdf?file=1&type=node&id=77.

Stiglitz, J. E.1989. Markets, Market Failures, and Development, *American Economic Review* 79(2): 197–203.

Stokes, L. C. 2013. The politics of renewable energy policies: The case of feed-in tariffs in Ontario, Canada. *Energy Policy* 56: 490–500. Available online: https://doi: 10.1016/j.enpol.2013.01.009.

Stadelmann, M. 2017. Mind the gap? Critically reviewing the energy efficiency gap with empirical evidence. *Energy Research and Social Science* 27: 117–128.

Stavins, R. N. 2007. Environmental economics. *National Bureau of Economic Research Working Paper Series, Working paper* 13574.

Stiess, M., Maghelli, N., Kapitein, L. C., Gomis-Ruth, S., Wilsch-Brauninger, M., Hoogenraad, C. C., Tolic-Norrelykke, I. M. & Bradke, F. 2010. Axon extension occurs independently of centrosomal microtubule nucleation. *Science* 327: 704–707. doi:10.1126/science.1182179.

Stiess, I., Zundel, S. & Deffner, J. 2009. Making the home consume less – putting energy efficiency on the refurbishment agenda. ECEEE 2009 Summer study. Act! Innovate! Deliver. *Proceedings of the ECEEE 2009*. Summer Study. Stockholm: European Council for an Energy Efficient Economy, 1821–1827.

Stigka, E. K., Paravantis, J. A. & Mihalakakou, G. K. 2014. Social acceptance of renewable energy sources: A review of contingent valuation applications. *Renewable and Sustainable Energy Reviews* 32, 100–106.

Streimikiene, D. & Alisauskaite-Seskiene, I. 2014. External costs of electricity generation options in Lithuania. *Renewable Energy* 64: 215–224.

Streimikiene, D. & Balezentis, T. 2013. Multi-criteria assessment of small scale CHP technologies in buildings. *Renewable and Sustainable Energy Reviews* 26: 183–189.

Streimikiene, D. & Balezentis, A. 2014. Assessment of willingness to pay for renewables in Lithuanian households. *Clean Technologies and Environmental Policy* 17(2): 515–531.

Streimikiene, D.; Balezentis, T. 2020. Willingness to Pay for Renovation of Multi-Flat Buildings and to Share the Costs of Renovation. *Energies 13*, 2721. Available online: https://doi.org/10.3390/en13112721

Streimikiene, D.; Balezentis, T.; Alebaite, I. 2020. Climate Change Mitigation in Households between Market Failures and Psychological Barriers. *Energies 13*, 2797. Available online: https://doi.org/10.3390/en13112797

Streimikiene D., Balezentis A., Alisauskaite-Seskiene I., Stankuniene G., Simanavicene Z. 2019. A Review of Willingness to Pay Studies for Climate Change Mitigation in the Energy Sector. *Energies* 12(8): 1–33. Available online: https://doi.org/10.3390/en12081481.

Streimikienė, D. & Mikalauskiene, A. 2014. Lithuanian consumer's willingness to pay and feed-in prices for renewable electricity. *Amfiteatru Economic Journal* 16(36): 594–605.

Su W., Liu M., Zeng Sch., Štreimikienė D., Baležentis T., Ališauskaitė-Šeškienė I. 2018. Valuating renewable microgeneration technologies in Lithuanian households: A study on willingness to pay. *Journal of Cleaner Production* 191: 318–329.

Sun, C., Yuan, X. & Xu, M. 2015. The public perceptions and willingness to pay: From the perspective of the smog crisis in China. *Journal of Cleaner Production* 112(2). Available online: https://doi: 10.1016/j.jclepro.2015.04.121.

Sun, P., & Nie, P. 2015. A comparative study of feed-in tariff and renewable portfolio standard policy in renewable energy industry. *Renewable Energy* 74: 255–262. Available online: https://doi: 10.1016/j.renene.2014.08.027.

Sundt, S. & Rehdanz, K. 2015. Consumers' willingness to pay for green electricity: A meta-analysis of the literature. *Energy Economics* 51: 1–8.

Tietenberg, T. 2003 The tradable –permit approach to protecting the commons: Lessons for climate change. *Oxford Review of Economic Policy* 19(3): 400–419.

Tol, R.S.J. 2013. Targets for global climate policy: an overview. *Journal of. Econonomic Dynamics and Control* 37(5): 911–928.

Train, K. E. 2009. *Discrete choice methods with simulation.* Cambridge University Press, 385 p.

Uihlein, A. & Eder, P. 2009. Towards additional policies to improve the environmental performance of buildings. European Commission Joint Research Centre, Institute for Prospective Technological Studies. *JRC Scientific and Technical Reports* EUR 24015 EN. Available online: https://doi: 10.2791/29052.

United Nations Environmental Programme Sustainable Buildings & Climate Initiative (UNEP-SBCI) 2009. *Buildings and climate change.* Summary for decision-makers, Paris. Available online: www.greeningtheblue.org/sites/default/files/Buildings%20and%20climate%20change_0.pdf.

Urge-Vorsatz, D., Koppel, S., Liang, C., Kiss, B., Nair, G., Celikyilmaz, G. 2007. An Assessment of Energy Service Companies Worldwide. Available online: www.worldenergy. org/documents/esco_synthesis.pdf.

Van der Veen, Reinier AC & De Vries, L. J. 2009. The impact of microgeneration upon the Dutch balancing market. *Energy Policy* 37(7): 2788–2797.

Van Putten, M., Lijesen, M., Özel, T., Vink, N. & Wevers, H. 2014. Valuing the preferences for micro-generation of renewables by househoulds. *Energy* 71: 596–604.

Vecchiato, D. & Tempesta, T. 2015. Public preferences for electricity contracts including renewable energy: A marketing analysis with choice experiments. *Energy* 88: 168–179.

Vainio, T. 2011 Building renovation – a new industry? *Management and Innovation for a Sustainable Built Environment*, 20–23 June 2011, Amsterdam, Netherlands.

Viscusi, W. and Zeckhauser, R. 2006. The perception and valuation of the risks of climate change: a rational and behavioral blend. *Climatic Change* 77: 151–177.

Ward, D.O., Clark, C.D., Jensen, K.L., Yen, S.T., Russell, C.S., 2011. Factors influencing willingness-to-pay for the Energy STAR label. *Energy Policy* 39: 1450–1458.

Watson, J. 2004. Co-provision in sustainable energy systems: The case of microgeneration. *Energy Policy* 32(17): 1981–1990.

WEC (World Energy Council), 2004. *Energy Efficiency: A Worldwide Review.* London, UK.

WEC (World Energy Council), 2008. *Energy Efficiency Policies around the World: Review and Evaluation.* London, UK.

Weinsziehr, T., Grossmann, K., Groger, M., Bruckner, T. 2017. Building retrofit in shrinking and ageing cities: a case-based investigation. *Buildings research and information* 45 (3): 278–292.

Wheeler, S. M. 2008. State and municipal climate change plans: The first generation. *Journal of the American Planning Association* 74(4): 481–496, DOI: 10.1080/01944360802377973

Willis, K., Scarpa, R., Gilroy, R. & Hamza, N. 2011. Renewable energy adoption in an ageing population: Heterogeneity in preferences for micro-generation technology adoption. *Energy Policy* 39(10): 6021–6029.

Wilson,C., Dowlatabadi, H. 2007. Models of Decision Making and Residential Energy Use. *Annual Review of Environment and Resources* 32: 169–203.

Wiser, R. 2007. Using contingent valuation to explore willingness to pay for renewable energy: a comparison of collective and voluntary payment vehicles. *Ecological Economics* 62: 419–432.

Wittman, T., Morrison, Richter, J. & Bruckner, T. 2006. *A bounded rationality model of private energy investment decisions.* Berlin: Institute for Energy Engineering, Technical University of Berlin, 20 p.

Wood, L. L., Kenyon, A. E., Desvousges, W. H. & Morander, L. K. 1995. How much are customers willing to pay for improvements in health and environmental quality? *Electricity Journal* 8(4): 70–77.

Würtenberger, L., Bleyl, J. W., Menkveld, M., Vethman, P., & van Tilburg, X. 2012. Business models for renewable energy in the built environment. Renewable Energy Technology Deployment (IEA-RETD)/ECN-E—12-014. Available online: http://iea-retd.org/wp-content/uploads/2012/04/RE-BIZZ-final-report.pdf.

Wüstenhagen, R., Wolsink, M. & Bürer, M. J. 2007. Social acceptance of renewable energy innovation: An introduction to the concept. *Energy Policy* 35(5), 2683–2691.

Yamamoto, Y. 2015. Opinion leadership and willingness to pay for residential photovoltaic systems. *Energy Policy* 83: 185–192.

Yaping, H., Oliphant, M. & Hu, E. J. 2016. Development of renewable energy in Australia and China: A comparison of policies and status. *Renewable Energy* 85(C): 1044–1051.

Yoeli E. 2010 Does social approval stimulate prosocial behavior? *Evidence from a field experiment in the residential electricity market.* Ann Arbor, MI, USA: ProQuest LLC.

Yue, T., Long, R., Chen, H. 2013. Factors influencing energy-saving behavior of urban households in Jiangsu Province. *Energy Policy* 62: 665–675.

Zhang Ch., Wang Q., Zeng Sh., Baležentis T., Štreimikienė D., Ališauskaitė-Šeškienė I., Chen X. 2019. Probabilistic multi-criteria assessment of renewable microgeneration technologies in households. *Journal of Cleaner Production* 212: 582–592.

Zeng, S., Jiang, C., Ma, C., & Su, B. 2018. Investment efficiency of the new energy industry in China. *Energy Economics* 70: 536–544. Available online: https://doi: 10.1016/j.eneco.2017.12.023.

Zhang, H., Li, L., Zhou, D., & Zhou P. 2014. Political connections, government subsidies and firm financial performance: Evidence from renewable energy manufacturing in China. *Renewable Energy* 63: 330–336. Available online: https://doi: 10.1016/j.renene.2013.09.029.

Zhao, Z., Chang R., & Chen Y. 2016. What hinder the further development of wind power in China?—A socio-technical barrier study. *Energy Policy* 88: 465V476. Available online: https://doi: 0.1016/j.enpol.2015.11.004.

Zhou, H., & Bukenya, J.O., 2016. Information inefficiency and willingness-to-pay for energy-efficient technology: A stated preference approach for China Energy Label. *Energy Policy* 91: 12–21.

Zografakis, N., Sifaki, E., Pagalou, M., Nikitaki, G., Psarakis, V. & Tsagarakis, K. P. 2010. Assessment of public acceptance and willingness to pay for renewable energy sources in Crete. *Renewable and Sustainable Energy Reviews* 14(3): 1088–1095.

Zoric, J., Filippini, M., & Hrovatin, N. 2012. Determinants of Energy-Efficient Renovation Decisions of Slovenian Homeowners. *Presentation at the IAEE Conference*, Venice.

Zorić, J. & Hrovatin, N. 2012. Household willingness to pay for green electricity in Slovenia. *Energy Policy* 47: 180–187.

Zyadin, A., Halder, P., Kähkönen, T., & Puhakka, A. 2014. Challenges to renewable energy: A bulletin of perceptions from international academic arena. *Renewable Energy* 69: 82–88. Available online: https://doi: 10.1016/j.renene.2014.03.029/

Zundel, S., Stier, I. 2011 Beyond profitability of energy-saving measures—attitudes towards energy saving, *Journal of Consumer Policy* 34: 91–105.

Index

renewable energy, 5, 13, 17, 25, 53, 59,
   93, 140–141; communities, 63–64;
   development programmes, 89;
   infrastructure, 94, 96; installations,
   113; policies, 143; producers, 105;
   projects, 25, 94–97; promotion
   measures, 94; resources, 57, 120;
   sources, 34, 54–56, 58, 83, 98–100,
   102–104, 139–141, 143–144; standards,
   97; systems, 114; targets, 55, 97, 101,
   120, 143; technologies, 20–21, 32–33,
   38, 61, 93, 96, 114, 120, 143
renovation activities, 73, 124, 127
renovation projects, 71–74, 80, 119
resistant innovations, 21

security of energy supply, 3–4
sellers, 84
single-choice experiment, 32
social benefits, 2, 11–12, 22, 38
society, 3, 5, 12, 84, 86, 92, 144
solar collectors, 61, 93, 113, 142
solar PV, 32, 50; technology, 37
stated preference, 15, 22; method, 14–15;
   studies, 17; techniques, 15
State-funded programmes, 89
stirling engines, 20
strategic documents, 54, 64, 66, 80–81
structural barriers, 9, 106–107
subsidy, 76–77, 119–120; programme,
   114
suppliers, 21, 65–66, 75, 109, 131–132

sustainability, 74
systematic deviation, 11

taxes, 4–7, 11, 66, 69, 79, 96, 99, 111–112,
   115–116, 119, 131, 143–144
technology barriers, 126
The European Union's (EU) Climate and
   Energy Policy Strategy, 53
The National Strategy for Climate
   Change Management Policy of
   Lithuania, 64
thermostatic regulators, 61
traditional economy, 6
transaction, 36, 86; costs, 87, 107–108,
   115, 123, 126–127

unemployment, 83, 86

valuation methods, 14, 16
ventilation systems, 71, 74
voluntary measures, 106, 112, 117,
   119–121
voluntary participation, 16
vulnerability, 128

warm season, 42–43, 46–47
welfare, 21, 30–32, 37, 92, 142; costs, 135,
   137; economics, 135, 137; losses, 136
wind farms, 16
wood, 20, 78; chips, 20

zero-energy buildings, 64, 114